Maximize Your Lab Time!

Access included with every new book.

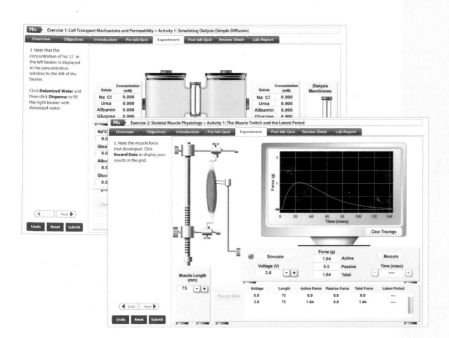

Access Online Labs

REGISTER NOW!

Registration will let you:

Perform 63 lab simulations at your own pace

Experiment with multiple variables

Review basic concepts

Assess your learning with Pre-Lab and Post-Lab Quizzes and Review Sheet Questions

Save all of your results as a PDF Lab Report

www.physioex.com

TO REGISTER

1. Go to www.physioex.com
2. Select the cover for *Experiments in Physiology*.
3. Click "Register."
4. Follow the on-screen instructions and scratch off the silver foil coating below to reveal your pre-assigned access code to create your account.

Your Access Code is:

TO LOG IN

1. Go to www.physioex.com
2. Select the cover for *Experiments in Physiology*.
3. Enter your Login Name and Password.
4. Click "Log In."

Hint:
Remember to bookmark the site after you log in.

Technical Support:
http://247pearsoned.custhelp.com

This access code may only be used by the original purchaser.

peel here

peel here

If there is no silver foil covering the access code, it may already have been redeemed, and therefore may no longer be valid. In that case, you can purchase access online using a major credit card or PayPal account. To purchase access online, go to www.physioex.com, click on the cover for Experiments in Physiology, select "Buy Now," and follow the on-screen instructions.

ELEVENTH EDITION

Experiments in Physiology

David A. Woodman
University of Nebraska, Lincoln

Gerald D. Tharp
University of Nebraska, Lincoln

PEARSON

Boston Columbus Indianapolis New York San Francisco Upper Saddle River
Amsterdam Cape Town Dubai London Madrid Milan Munich Paris Montréal Toronto
Delhi Mexico City São Paulo Sydney Hong Kong Seoul Singapore Taipei Tokyo

Senior Acquisitions Editor: Kelsey Churchman
Director of Development: Barbara Yien
Assistant Editor: Ashley Williams
Project Manager Team Lead: Nancy Tabor
Project Manager: Jessica Picone
Production Management and Composition: Integra
Design Manager: Marilyn Perry
Senior Manufacturing Buyer: Stacey Weinberger
Senior Marketing Manager: Allison Rona
Cover Photo Credit: Science Picture Co./Corbis

**Copyright © 2015, 2011, 2008 by Pearson Education, Inc., 1301 Sansome St.,
San Francisco, CA 94111.** All rights reserved. Manufactured in the United States of America.
This publication is protected by copyright, and permission should be obtained from the
publisher prior to any prohibited reproduction, storage in a retrieval system, or transmission
in any form or by any means, electronic, mechanical, photocopying, recording, or likewise.
To obtain permission(s) to use material from this work, please submit a written request to
Pearson Education, Inc., Permissions Department, 1900 E. Lake Ave., Glenview, IL 60025.
For information regarding permissions, call (847) 486-2635.

Many of the designations used by manufacturers and sellers to distinguish their products are
claimed as trademarks. Where those designations appear in this book, and the publisher was
aware of a trademark claim, the designations have been printed in initial caps or all caps.

Safety Notification
The Author and Publisher believe that the lab experiments described in this publication,
when conducted in conformity with the safety precautions described herein and according
to the school's laboratory safety procedures, are reasonably safe for the student to whom
this manual is directed. Nonetheless, many of the described experiments are accompanied
by some degree of risk, including human error, the failure or misuses of laboratory or
electrical equipment, mis-measurement, chemical spills, and exposure to sharp objects,
heat, bodily fluids, blood, or other biologics. The Author and Publisher disclaim any liability
arising from such risks in connection with any of the experiments contained in this manual.
If students have any questions or problems with materials, procedures, or instructions on
any experiment, they should always ask their instructor for help before proceeding.

Library of Congress Cataloging-in-Publication Data

Tharp, Gerald D.
     Experiments in physiology/Gerald D. Tharp, David A. Woodman.—Eleventh edition.
        pages cm
     ISBN 13: 978-0-321-95773-3
     ISBN 10: 0-321-95773-3
     1. Physiology, Experimental—Laboratory manuals.   I. Woodman, David A.   II. Title.
QP44.T53 2015
612.0078—dc23

                                                                    2014001812

19 2019

www.pearsonhighered.com   ISBN 13: 978-0-321-95773-3
                                 ISBN 10:   0-321-95773-3

Contents

Preface

To the Instructor

A laboratory manual is never the work of one author alone; it represents a blend of ideas from other lab manuals, other teachers, and personal experience in the laboratory. Having taught physiology and directed the lab program for many years, we have been inspired to create and continue to improve a lab manual that will best serve our students' learning of the material. We have selected the experiments in this manual because they fulfill two key criteria:

1. They produce consistently successful results—students need not be trained scientists to get meaningful data.
2. They teach significant physiological concepts.

Hallmark Features

The following hallmark features of the manual make it an effective teaching and learning tool:

- **90 wide-ranging exercises, organized into 22** "teaching units" are suitable for 2- or 3-hour laboratory periods in one-semester introductory courses or upper-level one- or two-semester courses. Experiments for both humans and animals ranging from simple to complex give instructors the flexibility to select activities appropriate for their course.
- **Access to the PhysioEx™ 9.1 website,** featuring 63 online lab simulations organized into 12 exercises, is automatically included at no additional fee with each new copy. Instructors can use PhysioEx™ simulations to replace wet labs that are too expensive or time-consuming or that require unavailable materials, and students can repeat labs as often as they like. Look for the references to applicable PhysioEx™ experiments at the beginning of the chapters.
- **Integration of modern technology,** in addition to the use of physiograph-based equipment—exercises that use Vernier®, PowerLab®, and BIOPAC® data acquisition systems—appears throughout.
- **Laboratory reports that engage students in higher-level thinking** using data tables, graphs, and critical thinking questions help students relate their findings to the key concepts discussed in lecture.
- **A comprehensive Instructor Guide**—available online—provides sample data, graphs, and answers to questions in the lab reports plus lists of materials and equipment needed for each teaching unit and the quantities needed for a lab of 20 to 24 students.

What's New to This Edition?

The overall goals for the eleventh edition have been to expand the background information that helps students prepare for lab exercises, relate lab activities to real-world applications, enhance critical thinking features, and update the use of technology. Students will benefit from a variety of new content and features in the eleventh edition.

- **Inquiry-based activities** have been added to encourage lab students to think like scientists.
- **"Explain This" questions** that focus on interpretation and analysis throughout a procedure have been added.
- **The new Appendix on Inquiry-Based Activities** provides information on interpreting graphs and using computers to graph.
- The eleventh edition is **now integrated with PhysioEx™ 9.1.** PhysioEx™ 9.1 lab simulations are new free-form experiments that interact with students on a deeper level within the scientific process.
- **More Clinical Application feature boxes** show students the relevance of what they're learning and how it relates to clinical situations, diseases, and medical devices.
- Frequent **"Stop and Think" features** prompt students to consider key questions before moving on to the next step in an exercise. "Stop and Think" encourages critical thinking and observation skills, which are essential for successfully completing the Laboratory Report.

To the Student

The study of physiology is only half accomplished if you never enter the laboratory. It is one thing to hear a concept explained in lecture but quite another to see the concept unfold before your eyes in a laboratory experiment. The study of physiology is fascinating and practical—fascinating for its examination of the awesome complexity of process and practical for its future usefulness in our lives.

The experiments presented in this manual are designed to illustrate the basic principles of physiology. They are also meant to develop your ability to carry out measurements, observe, and formulate reasonable deductions—characteristics of the scientific process.

Physiology might appear at first to be an easy science to master, especially to students whose prior schooling has included some study of the heart, brain, eye, and ear. As the complexity of the subject becomes apparent, however, some students might become discouraged. We urge you to accept the difficulties as a challenge; you will find that the more effort you expend, the more interesting physiology will become.

The following suggestions will make the laboratory experience more valuable to you.

1. Study the lab experiment and the lab report *before* coming to the lab. Usually, your instructor will introduce the lab, but this introduction is to help you organize your work, not to give all the details for conducting the experiments.

2. Become acquainted with the location of equipment and supplies. Use the instructor's introductory comments to help you plan how your team can accomplish the lab work most effectively.

3. Participate actively in the lab. Do not expect to listen passively and let others do the work. Research has shown that true learning occurs best when a person learns actively and that active learning is stored longer in the memory systems of the brain than passive learning.

4. Be prepared to work closely with others. Physiology lab work is a team effort that requires an exchange of information and interaction with your classmates and with your instructor. Working closely with others is an important feature of the lab experience, one that will provide benefits to you beyond the mere acquisition of knowledge.

5. As you conduct the experiments, try to relate the theoretical information presented in your lecture and textbook with your lab observations. Don't just perform the work mechanically. There is always the danger that, as you struggle with technical difficulties, you will lose sight of the purpose of the experiment. You should continually ask yourself, "What is this experiment trying to show us?"

6. Get in the habit of promptly recording all data as soon as they become available. Much information is lost if you don't write things down immediately. If a recording is made of some parameter, write on the record the date, experiment, experimental conditions, and results so that you have complete data when you examine the record later.

7. Use the lab report as a guide for recording the data from the experiments and studying the major concepts explored in each lab. If used properly (that is, to engage your mind in active, critical thought about what you have seen in lab), the lab report can be a useful learning device. If you don't do your own reports, you will have missed a valuable opportunity to put your mind to work and learn something.

8. Ask for help when you don't understand how something works. Your instructor will gladly help you get your experiment working, but you should also attempt to solve minor problems yourself. Check the equipment to make sure that the connections are tight, or adjust the settings of the data-collecting units. Don't be too discouraged when an experiment fails or you obtain data that do not agree with the expected results. Because of "biological variation," there will be times when things don't work out exactly right (this also happens in real life), but try not to get discouraged, and keep doing your best.

9. When the lab is completed, clean your lab table and discard any waste in the wastebasket. Return all equipment and supplies to their appropriate places.

10. **CAUTION!** Some experiments in human physiology in this manual, such as the step tests and the exercises to be done in Chapters 21 and 22, induce some degree of cardiovascular stress. *Students who have cardiovascular difficulties, such as cardiovascular insufficiency or hypertension, should not take part in any experiment that causes cardiovascular stress unless they have permission from their physician.* If you feel that you should not participate in any experiment for personal health reasons, be sure to tell your instructor. If you suffer any ill effects during an experiment, stop and inform your instructor.

If you keep these suggestions in mind and approach the physiology lab with a positive attitude, you will be richly rewarded with a positive learning experience.

Good luck. We hope your study of physiology is as exciting as ours has been.

Acknowledgments

I continue to be enthusiastic about my involvement in *Experiments in Physiology*. My long-standing association with Gerald Tharp has been fruitful. He has been an encouraging colleague and a model teacher whose dedication and abilities have served me well. I would also like to thank those who have made this possible: my wife Rosemary, whose love encouraged my efforts; my children Marcus and Sarah, who make every moment with them special; and my parents, who financially and emotionally supported my educational goals, even though it meant that I had to move abroad.

I would like to thank those thousands of students, past and present, who have contributed to my growth as a teacher. Special thanks are necessary for those select few who became teaching assistants. Their intelligence and hard work made laboratory instruction a pleasure.

Finally, I would like to thank the editorial staff at Pearson Education, especially Ashley Williams; their expertise and patience has made this task very enjoyable.

DAVID A. WOODMAN

I would like to express my appreciation to many of my students for their constructive criticisms, which have improved the effectiveness of this manual. I am most indebted to my teaching assistants, who have worked closely with me in preparing, teaching, and criticizing the labs.

A major factor in any endeavor such as this is the support of those who make all of this effort worthwhile: my wife Dee, who has loved me in spite of the long hours I spend with my nose in the books; my children, Danny, Kathy, Tim, and Jeanine, who have given my life much meaning and excitement; and my parents, Evae and Glenn, who instilled in me the idea that each person is valuable in God's eyes and gave me the confidence to accomplish my goals in life.

Also, I would like to thank my first and best physiology professor, Paul Landolt, who made physiology so interesting that I decided to enter the field and follow in his footsteps, a decision I have never regretted.

I am very grateful to my co-author David A. Woodman for embracing the continuing challenge of revising this lab manual for the eleventh edition. David has taught physiology and has directed the physiology laboratory program at the University of Nebraska–Lincoln for over 20 years. He has made major contributions to our physiology and anatomy programs through his expertise in the use of computer programs and Internet applications to teaching. Thanks, Dave!

GERALD D. THARP

Reviewers

We are deeply grateful to the following reviewers who helped shape the direction of the eleventh edition:

Joslyn Ahlgren, University of Florida
Betsy C. Brantley, Valencia College
Tyson Chappell, Utah State University Eastern
Melody Danley, University of Kentucky
Paul Deeble, Mary Baldwin College
David Kurjiaka, Grand Valley State University
Cheryl L. Neudauer, Minneapolis Community & Technical College
Bradley Rabquer, Albion College
Gary Ritchison, Eastern Kentucky University
Chad M. Wayne, University of Houston–Downtown
Erica Wehrwein, Michigan State University
Heather Wilson-Ashworth, Utah Valley State University

Fundamental Physiological Principles

1

CHAPTER 1 INCLUDES:

 Vernier 1 Vernier® Activity   **PowerLab** 1 PowerLab® Activity

OBJECTIVES

After completing this exercise, you should be able to

1. Use proper units of measurement and understand the relationship between the SI (metric) system and the English customary system (FPS).
2. Understand and calculate the concentration of solutions, using alternate methods of measurement.
3. Define the terms *acid* and *base* and explain the purpose and usage of the pH scale.
4. Explain the importance of buffers in biological systems.
5. Demonstrate the effects of buffers by using a pH measuring system.

Physiology is a quantitative science; physiologists are constantly trying to measure changes occurring in living organisms. Experimental work in physiology therefore requires knowledge of certain fundamental principles such as units of measurements, concentration of solutions, and acid–base balance. This exercise is designed to acquaint or reacquaint students with these principles, which we will use during our study of physiology. For students with previous courses in biology and chemistry, this will be old material. For others, it will be a first encounter with these concepts. It is hoped that all students will be on more even ground after completing this lab.

Units of Measurement

The following units are used fairly frequently in physiology:

Length

kilometer (km) = 1,000 m = 0.62 mi
1 mi = 1.61 km = 1,760 yd
hectometer (hm) = 100 m

dekameter (dkm) = 10 m
meter (m) = 39.37 in. = 1.09 yd
decimeter (dm) = $\frac{1}{10}$ m = 10 cm
centimeter (cm) = $\frac{1}{100}$ m = 10 mm = 0.3937 in.
1 in. = 2.54 cm
millimeter (mm) = $\frac{1}{1000}$ m = 1,000 μm
    (microns) = $\frac{1}{25}$ in.
micrometer (μm) = one-millionth m
   = 1,000 nm = 1 micron
nanometer (nm) = one-billionth m
   = 1 mμm = 10 Å
angstrom (Å) = 0.1 nm
picometer (pm) = one-trillionth m

Volume

liter (L) = 1,000 ml = 1,000 cc = 1.05 qt = 0.264 gal
deciliter (dl) = 0.1 L = 100 ml
milliliter (ml) = $\frac{1}{1000}$ L = 1,000 μl
ounce (oz) (fluid) = 8 fl drams = 29.57 ml
quart (qt) (fluid) = 32 oz = 946 ml = 0.946 L

Weight

metric ton = 1,000,000 g = 2204.62 lb
kilogram (kg) = 1,000 g = 2.2 lb = 35.27 oz
gram (g) = 1,000 mg
454 g = 1 lb = 16 oz
1 oz = 28.35 g
milligram (mg) = $\frac{1}{1000}$ g = 1,000 μg
microgram (μg) = $\frac{1}{1000}$ mg = 1 gamma

Temperature

0° centigrade (°C) = 32° Fahrenheit (°F) = 273 Kelvin (K)
°C = $\frac{5}{9}$ (°F −32)
°F = $\frac{9}{5}$ °C + 32

Copyright © 2015, 2011, 2008 Pearson Education, Inc.

Pressure

Pressure is force per unit of area.

1 atmosphere = 34.0 ft of water = 760 mm
(or 29.92 in.) of mercury (Hg) = 14.7 lb/in.^{2}

Energy

One calorie (cal) is the amount of energy
required to heat 1 g of water 1°C (at 15°C).

One kilocalorie (large calorie or kcal) is the
amount of energy required to heat 1 kg of
water 1°C (at 15°C).

1 kcal = 1,000 cal = 3,086 ft-lb = 426.4 kg-m

1 g of carbohydrate = 4.1 kcal

1 L of oxygen used in burning glycogen
(RQ of 1) = 5.047 kcal = 15,575 ft-lb =
2,153 kg-m [RQ = (volume CO_2
produced)/(volume O_2 consumed)
during metabolism]

1 L of oxygen in a closed circuit system =
4.825 kcal in the post absorptive state (RQ
assumed to be 0.82) and 4.862 kcal on
an ordinary mixed diet (RQ assumed to
be 0.85)

Work

Work is force times the distance through which
it acts.

1 ft-lb = 1 lb of force times 1 ft

1 kg-m = 7.23 ft-lb = 0.002343 kcal =
2.343 g-kcal

1 kg-m = 1 kg of force times 1 m

Power

Power is work or energy per unit of time.

1 horsepower (HP) = 33,000 ft-lb/min = 550
ft-lb/sec = 4,564 kg-m/min = 76.07 kg-m/sec =
746 watts (W) = 10.694 kcal/min =
0.178 kcal/sec

1 kilowatt (kw) = 1,000 W = 1.341 HP =
0.239 kcal/sec

Concentration of Solutions

Many of the physiological properties of solutions
depend on the number of molecules, ions, or particles in the solution; therefore, it is important that
you understand the various means of expressing
concentrations used in physiology.

Percentage (%) Solutions

This is probably the simplest means of expressing concentration and one that is commonly used.
Percent means "parts in 100." Percentage is the
number of grams of solute dissolved in 100 ml
(1 deciliter) of solution. It is calculated using this
formula:

$$\text{Percentage} = \frac{\text{Grams of solute}}{\text{Volume of solution}} \times 100$$

Thus, a 12% solution **weight to volume (W/V)**
of glucose would contain 12 g of glucose in each
100 ml of solution (12 g/dl), or 120 g/L. If 2 g of
NaCl is dissolved in 25 ml of water, the percentage
will be:

$$\text{Percentage} = \frac{2 \text{ g}}{25 \text{ ml}} \times 100 = 8\%$$

In living organisms, the concentration of
many substances is so low that it is more easily
expressed as **milligrams/deciliter (mg/dl)**. For
instance, the average blood glucose concentration is approximately 90 mg/dl. This simply means
that in every 100 ml of blood, there is 90 mg of
glucose. If this concentration were expressed as
percentage, it would be 0.09%, which is a more
awkward number to use.

Molar (M) Solutions

A 1 molar (M) solution contains 1 mole of solute
in 1 L of solution. One mole of a substance contains 6.024×10^{23} molecules (Avogadro's number).
Thus, solutions of equal molarity have the same
number of molecules in solution, even though
their molecular weights might be different. One
mole is equal to the molecular weight (MW) or
atomic weight of the solute in grams.

For example, the MW of glucose is 180. To prepare a 1 M solution of glucose, we would weigh
out 180 g of glucose and dissolve it in a total volume of solution (solvent + solute) of 1 L. Eighteen
grams of glucose in 100 ml of solution would also
be a 1 M and 18% solution. Why?

To make a 1 M solution of NaCl (58.5 MW), we
would dissolve 58.5 g of NaCl in 1 L of solution. This
would be the same as 5.85 g of NaCl in 100 ml, or
a 5.85% solution. You can see from these examples
that decreasing the amount of solute and solution by
the same proportion does not change the concentration of the solution.

Because of the low concentrations of solutes in
body fluids, we often use millimolar (mM) measurements in physiology. If 180 mg of glucose is

Copyright © 2015, 2011, 2008 Pearson Education, Inc.

dissolved in 1 L of solution, a 1 mM concentration is produced.

Glucose concentration in the blood = 90 mg/100 mL = 900 mg/L.

$$\frac{900 \text{ mg}}{180 \text{ mg/mM}} = 5 \text{ mM glucose}$$

Osmolar (Osm) Solutions

Osmolar concentrations are used mainly in the biological sciences to express the osmotic effect of a solution. To understand their use, we will look at some examples. (Osmosis and osmotic effects are further explained in Chapter 2, "Movement Through Membranes.")

A membrane permeable only to water separates a container into two compartments (Figure 1.1a). Water molecules are in side A, and glucose molecules are trapped in side B. In an effort to reach equilibrium, the water molecules pass through the membrane into side B, moving from a higher to a lower water concentration. We call this water movement **osmosis**. The force of the water movement on the membrane is called the **osmotic pressure** and is determined by the number of molecules in side B that cannot penetrate the membrane. The 1 M glucose solution is 1 Osm in the osmotic pressure it produces on the membrane.

If the concentration of glucose (a nonelectrolyte) is doubled in side B to 2 M (Figure 1.1b), the water moves across with twice the osmotic pressure because there are twice as many osmotically active particles in solution. Thus, the 2 M glucose solution is 2 Osm in its osmotic effect.

Some electrolyte molecules, such as NaCl (table salt), do not remain as molecules in solution but dissociate into ions.

$$1 \text{ mole NaCl} \rightarrow 1 \text{ mole Na}^+ + 1 \text{ mole Cl}^-$$

Each ion acts as an osmotically active particle to cause water movement into the region of higher solute concentration (osmosis). Thus, a 1 M NaCl solution produces the same osmotic pressure as a 2 M glucose solution (Figure 1.1c). Both solutions are therefore 2 Osm in their osmotic concentration. The osmole provides a measure of a solution's ability to produce osmosis or osmotic pressure.

The osmotic particles in the cells of a mammal have a concentration of 0.3 Osm. If we bathe the cell in a solution having the same osmolar concentration (e.g., 0.3 M glucose or 0.15 M NaCl), there will be no net movement of water in or out of

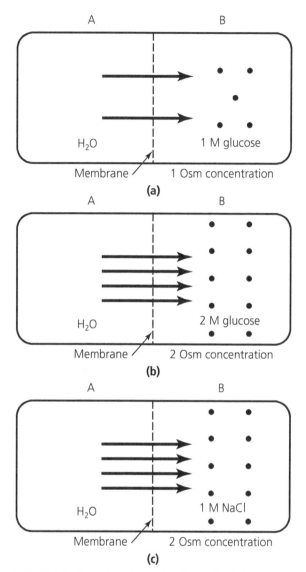

FIGURE 1.1   Examples of osmotic effects of solutions.

the cell, and the cell will retain its size and shape (Figure 1.2). Solutions having the same osmolar concentration as the concentration inside the cell are said to be **isotonic** (same "tone" or "tension"). Solutions with a higher osmolar concentration are called **hypertonic**, and those with a lower osmolar concentration are **hypotonic**. Which way will water move if a cell is placed in each of these solutions? Some examples of osmolar calculations are as follows:

To make a 1 Osm solution of NaCl (58.5 MW), dissolve 58.5 g/2 = 29.25 g in each liter of solution (two ions formed in solution).

To make a 1 Osm solution of $CaCl_2$ (110 MW), dissolve 110 g/3 = 36.6 g in each liter because $CaCl_2$ dissociates into three ions in solution.

Copyright © 2015, 2011, 2008 Pearson Education, Inc.

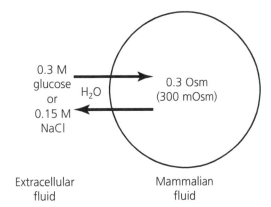

FIGURE 1.2   Example of osmotic equilibrium.

Equivalent (Eq) or Milliequivalent (mEq) Solutions

This means of expressing the concentration of ions is used extensively in chemistry and the various biological sciences. We will not use it in this manual, but you should be aware of it because it appears often in textbooks and research papers. Equivalent weights reflect the combining power of substances during a chemical reaction, which in turn depends on the valence (charge) of the atoms involved. For univalent ions such as Na^+, K^+, and Cl^-, the milliequivalents are equal to millimoles. Only when we deal with divalent (Ca^{2+}) or trivalent (Fe^{3+}) ions will the milliequivalent and millimole concentrations differ.

▶ ACTIVITY 1.1

Concentration of Solutions

To solve concentration problems, we need to know the relationships between percentage, molarity, and osmolarity concentrations. If these are related in a stepwise fashion, it is easier to understand and solve such problems. As an example, let us use a series of steps to solve the following problem: What is the osmolar concentration of a 2% solution of KCl?

1. Percentage refers to the number of grams in 100 ml of solution.

   2% KCl = 2 g of KCl dissolved in 100 ml of solution

2. To go from percentage to molarity, we must first determine the number of grams in 1 L (1,000 ml) of solution.

   2 g KCl in 100 ml = 20 g KCl in 1,000 ml (1 L) (because 1,000 is 10 × 100)

3. To determine molarity, we first determine the number of grams per liter in a 1 M solution of the substance.

   1 M KCl (74 MW) = 74 g/L

   Therefore, 20 g/L of KCl is less than 1 M. Specifically, we have 20 g/74 g = 0.27 M of KCl.

4. To go from molarity to osmolarity, we need to know whether the substance is an electrolyte, which dissociates into ions in solution, or a nonelectrolyte, which remains as molecules in solution. KCl is an electrolyte that forms 2 ions (K^+ and Cl^-) in solution.

   Moles × Number of ions = Osmoles

   Therefore, 0.27 M × 2 ions = 0.54 Osm KCl. ◀

CLINICAL APPLICATION

Treating Severe Dehydration

Dehydration, a condition in which more water is lost from the body than is taken in, is a life-threatening condition usually brought about by conditions such as diarrhea, vomiting, disease, or too much exercise and heat exposure. As fluids are lost from the body, so are essential salts, preventing the body from functioning normally. Severe dehydration can cause rapid, weak pulse; fever; fast, deep breathing; or convulsions. Left untreated, it is fatal. Commercial oral rehydration fluids such as Pedialyte® or Enfalyte® contain carbohydrate, sodium, potassium, and chloride ions and a citrate with an osmolarity between 200 and 300 mOsm/L. These solutions can be administered to alleviate the symptoms of dehydration by restoring hydration levels and replenishing the lost electrolytes.

Acid–Base Balance

It is critical for the body's homeostasis (maintenance of a constant internal environment) that the concentration of hydrogen ions (H^+) in the blood be maintained within the narrow range of pH 7.0 to 7.8. If the pH goes below or above these limits, death results because most enzymes cannot operate properly if the pH is outside this range. In light of the great importance of H^+ regulation, let us review the concepts of pH and acid–base balance.

pH

The pH scale is simply a way of expressing small molar concentrations using whole numbers (Figure 1.3). It was devised by the Danish chemist

Copyright © 2015, 2011, 2008 Pearson Education, Inc.

									Neutral		Basic (Alkaline)							
H^+ conc.					Acid													
H^+ conc.	10^0	10^{-1}	10^{-2}	10^{-3}	10^{-4}	10^{-5}	10^{-6}	10^{-7}	10^{-8}	10^{-9}	10^{-10}	10^{-11}	10^{-12}	10^{-13}	10^{-14}			
pH	0	1	2	3	4	5	6	7	8	9	10	11	12	13	14			
OH^- conc.	10^{-14}	10^{-13}	10^{-12}	10^{-11}	10^{-10}	10^{-9}	10^{-8}	10^{-7}	10^{-6}	10^{-5}	10^{-4}	10^{-3}	10^{-2}	10^{-1}	10^0			

FIGURE 1.3   pH scale.

Soren Sorensen. Note in Figure 1.3 that a difference of one pH unit represents a 10-fold change in H^+ or OH^- concentration.

pH = Logarithm of reciprocal of hydrogen ion concentration

$$= \log \frac{1}{[H^+]}$$

For example:

$$pH = \log \frac{1}{1 \times 10^{-7} mol} = \log \frac{10^7}{1} = 7$$

Acids and Bases

Acid. Substance that dissociates into hydrogen ions (H^+).

Base. Substance that dissociates into hydroxyl ions (OH^-).

Salt. Substance that dissociates into neither H^+ nor OH^- in solution.

A major problem for the body is that many acids are produced during metabolism. One of the most important of these is carbonic acid, formed when carbon dioxide (CO_2) dissolves in body fluids:

$$CO_2 + H_2O \leftrightarrow H_2CO_3 \leftrightarrow H^+ + HCO_3^-$$

carbon water carbonic hydrogen bicarbonate
dioxide   acid   ion   ion

This is one of the most important chemical reactions you will encounter in your study of physiology. Other acids that lower the pH of body fluids are phosphoric, sulfuric, hydrochloric, lactic, keto, and fatty acids.

Buffer Systems

Even though the body produces tremendous quantities of acid each day, blood pH usually remains within the range of 7.35 to 7.45. This is made possible by the action of various buffer systems in the body fluids.

Buffer. Substance that prevents (resists) a drastic pH change when acids or bases are added to a solution.

Buffer mechanism. Mechanism that replaces strong acids or bases with weak acids or bases that produce less H^+ or OH^- in solution.

The major buffer systems found in most animals are the protein, phosphate, and bicarbonate systems. **Proteins** are the most abundant buffers, playing a critical role inside body cells as well as in the blood (e.g., albumins, hemoglobin). **Phosphates** are less abundant but perform important buffering in the intracellular fluid and kidney tubules.

Bicarbonates are most important in the buffering of the extracellular fluid (interstitial fluid and plasma). The bicarbonate system is a unique buffer because its components (HCO_3^- and CO_2) can be regulated by the renal and respiratory systems. This makes it a very powerful and flexible buffer, one that deserves closer scrutiny.

Bicarbonate   = H_2CO_3 + $NaHCO_3$
buffer   carbonic acid   sodium
                 (weak acid)   bicarbonate
                              (conjugate base)

Adding a strong acid or base to this system produces the following reactions:

HCl   + $NaHCO_3 \rightarrow H_2CO_3$ NaCl
hydrochloric acid   (weak acid)   (salt)
 (strong base)

NaOH   + $H_2CO_3 \rightarrow NaHCO_3$ + H_2O
sodium   (weak base)
hydroxide
(strong base)

Thus, the bicarbonate buffer chemicals replace strong acids and bases with weak acids and bases that dissociate only weakly and, therefore, produce little change in pH. The following activity demonstrates how buffers moderate the pH effects of acids and bases.

Copyright © 2015, 2011, 2008 Pearson Education, Inc.

▶ ACTIVITY 1.2

Acid–Base Balance

Materials

- ☐ pH meter and electrode
- ☐ Concentrated HCl (10 ml per group)
- ☐ Concentrated NaOH (10 ml per group)
- ☐ Distilled water (400 ml per group)
- ☐ Phosphate buffer (400 ml per group)

1. Allow the pH meter to warm up with the selector switch on the standby position. Set the temperature selector to room temperature.
2. Immerse the pH electrode in 150 ml of distilled water contained in a 250-ml beaker. Turn the selector switch to pH and record the pH of the water. Return the switch to standby.
3. Add one drop of concentrated hydrochloric acid (HCl) to the distilled water, mix with a stirring rod, and record the new pH. Repeat these procedures for four additional drops of HCl, recording the new pH after each drop is added. Be careful not to add the acid directly on the electrode because this could damage the electrode. Rinse the electrode with distilled water in a waste beaker before going on to the next step.
4. Add one drop of concentrated sodium hydroxide (NaOH) to a fresh 250-ml beaker of distilled water (150 ml), mix, and record the new pH. Repeat this with four additional drops of base, recording the new pH after each drop is added.
5. Remove the pH electrode from the solution, rinse with distilled water, and dry with lint-free paper.
6. Immerse the electrode in 150 ml of phosphate buffer solution in a 250-ml beaker and record the pH. This buffer is a mixture of dibasic sodium phosphate (Na_2HPO_4) and sodium acid phosphate (NaH_2PO_4).
7. Repeat the addition of concentrated acid and base, as in steps 3 and 4, to this buffer solution. Record the new pH after each drop is added and mixed.

1.2 VERNIER® VERSION

Materials (per group)

- ☐ 1 computer
- ☐ 2 Vernier® pH sensors
- ☐ 1 Vernier® LabPro
- ☐ Concentrated HCl (10 ml)
- ☐ Concentrated NaOH (10 ml)
- ☐ Distilled water (400 ml)
- ☐ Phosphate buffer (400 ml)

1. Connect pH sensors to Channels 1 and 2 on the Vernier® LabPro.
2. Connect the LabPro to an available USB port on your computer and turn on the computer and the LabPro.
3. Start the Logger Pro® software on the computer by opening the file named "03 Acids and Bases" from the *Biology with Vernier* folder of Logger Pro. You are now ready to collect data.
4. Rinse the probes thoroughly with distilled water and then place each in 150 ml of distilled water contained in a 250-ml beaker. Label one beaker "acidic" and the other "basic."
5. Click **Collect** to begin making measurements. When the readings are stable, click **Keep** and then type **0** in the edit box. Press **Enter**. Notice that values are automatically entered in the data table to the left.
6. Add one drop of concentrated hydrochloric acid (HCl) to the distilled water in the beaker labeled "acidic" and mix with a stirring rod. Be careful not to add the acid directly on the electrode because this could damage the electrode.
7. Add one drop of concentrated sodium hydroxide (NaOH) to the beaker of distilled water labeled "basic" and mix with another stirring rod. Once again, be careful not to add the base directly on the electrode because this could damage the electrode.
8. When the readings stabilize, click **Keep** and then type **1** in the edit box. Press **Enter**.
9. Repeat these procedures for four additional drops of HCl and NaOH, recording the new pH after each drop is added. Type in the total number of drops entered in the text box after each drop. It is important to add the same number of drops to each beaker or your results will be meaningless.
10. Click **Stop** when you have added all five drops. Move your data to a stored data run. Choose **Store latest run** from the **Experiment** menu.
11. Rinse the probes with distilled water and return them to the storage bottles.
12. Repeat these procedures, using a phosphate buffer solution instead of distilled water. Label the clean beakers as you did before.

Copyright © 2015, 2011, 2008 Pearson Education, Inc.

13. Immerse the electrodes in 150 ml of phosphate buffer solution in a 250-ml beaker and record the pH. This buffer is a mixture of dibasic sodium phosphate (Na_2HPO_4) and sodium acid phosphate (NaH_2PO_4). Repeat steps 5 to 10 to start recording your data.

14. Repeat the addition of concentrated acid and base, as in steps 5 to 7 and then step 10, on this buffer solution. Record the new pH after each drop is added and mixed. Remember to click **Stop** after you have added the acid and once again after you have added the base.

15. You will notice that a graph for your runs with (a) water in the beakers and (b) buffer in the beakers has been created as you proceed through the experiment.

16. You can print the graph by selecting **Print graph** from the **File** menu.

1.2 POWERLAB® VERSION

Materials (per group)

- ☐ 1 computer
- ☐ 1 Powerlab®
- ☐ 2 Powerlab® pH sensors
- ☐ 1 Powerlab® pH pod
- ☐ Concentrated HCl (10 ml)
- ☐ Concentrated NaOH (10 ml)
- ☐ Distilled water (400 ml)
- ☐ Phosphate buffer (400 ml)

1. Connect the cable from the pH Pod into the Input 1 Pod Port on the front of the PowerLab®.
2. Attach the BNC connector on the pH electrode to the socket on the rear panel of the pH Pod.

Use the following steps to calibrate the pH electrode.

a. Obtain a beaker to use as a waste container.
b. Place 30–40 ml of pH 4.01 buffer in a 100-ml beaker.
c. Place 30–40 ml of pH 10.01 buffer in another 100-ml beaker.
d. Remove the pH electrode tip from its beaker and rinse the tip with distilled water, using your waste beaker to catch the drips.
e. Place the pH electrode into the beaker of pH 4.01 buffer.
f. In Chart, click **Start**.

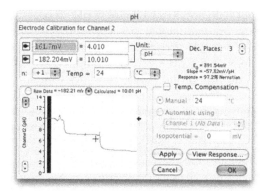

FIGURE 1.4 pH electrode calibration dialog box.

g. Record for 20 sec.
h. Remove the pH electrode from the buffer, rinse the tip in your waste beaker with distilled water, and then replace the tip into the pH 10.01 buffer.
i. Record for 30 sec.
j. Click **Stop**.
k. Rinse the pH electrode and return it to its beaker of pH 7.0 buffer.
l. Make a selection of your data in the pH channel.
m. From the pH channel function pop-up menu, choose **pH**. A dialog box will appear (Figure 1.4).
n. Select the data in the left-hand window that corresponds to pH 4.01 and click the arrow next to Point 1.
o. Select the data corresponding to pH 10.01 and click the arrow next to Point 2.
p. When you are done, click **OK** to return to the Chart view.

To do the experiment, follow the procedure below.

a. Immerse the pH electrode in 150 ml of distilled water contained in a 250-ml beaker.
b. In Chart, click **Start**.
c. Add one drop of concentrated hydrochloric acid (HCl) to the distilled water, mix with a stirring rod, and record the new pH after it stabilizes. Repeat these procedures for four additional drops of HCl, recording the new pH after each drop is added. Be careful not to add the acid directly on the electrode because this could damage the electrode. Click **Stop** after recording the pH. Pour distilled water over the electrode and into a waste beaker before going on to the next step.
d. Add one drop of concentrated sodium hydroxide (NaOH) to a fresh 250-ml beaker of distilled water (150 ml), mix, click **Start**

Copyright © 2015, 2011, 2008 Pearson Education, Inc.

in Chart, and record the new pH. Repeat this with four additional drops of base, recording the new pH after each drop is added.

e. Remove the pH electrode from the solution, rinse with distilled water, and dry with lint-free paper.

f. Immerse the electrode in 150 ml of phosphate buffer solution in a 250-ml beaker and record the pH. This buffer is a mixture of dibasic sodium phosphate (Na_2HPO_4) and sodium acid phosphate (NaH_2PO_4).

g. Repeat the addition of concentrated acid and base, as in steps 3 and 4, to this buffer solution. Record the new pH after each drop is added and mixed. ◄

CLINICAL APPLICATION

Acidosis

Acidosis is a condition where the body experiences increased acid levels, most often in the blood. It occurs when the arterial pH falls below 7.35, and it may be metabolic or respiratory in origin. While there are many different reasons for metabolic acidosis, it is typically caused by an increased production of metabolic acids or changes in the kidney's ability to excrete acid. Respiratory acidosis is usually caused by a build-up of CO_2 in the blood due to lowered ventilation rates. In either case, acidosis can be dangerous if untreated. The symptoms are dependent upon the specific type of acidosis; most will require medical intervention.

Copyright © 2015, 2011, 2008 Pearson Education, Inc.

Movement Through Membranes

CHAPTER 2 INCLUDES:

 2 Vernier® Activities

PowerLab 2 PowerLab® Activities

PhysioEx™ 9.I

For more exercises on Movement Through Membranes, visit PhysioEx™ (www.physioex.com) and choose Exercise 1: Cell Transport Mechanisms and Permeability.

⚠ CAUTION!

Parts of this chapter involve experiments with non-human mammalian blood. Disposable latex gloves must be worn, and the protocol in "Precautions for Handling Blood" must be followed. (See "General Safety in the Laboratory" inside the front cover of this text.)

OBJECTIVES

After completing this exercise, you should be able to

1. Define the term *diffusion* and examine the parameters that govern the movement of molecules.
2. Demonstrate and measure the effects of temperature on the rate of diffusion using a conductivity probe.
3. Examine the effect of concentration gradients on the rates of diffusion.
4. Define and understand the terms *osmosis* and *osmotic pressure*.
5. Define the term *tonicity* and understand the difference between the terms *isotonic*, *hypotonic*, and *hypertonic*.
6. Examine the effects of molecular size and lipid solubility on cell permeability.

Throughout the body, we find many types of membranes, such as the capillary membrane, alveolar membrane, cell membrane, and nuclear membrane. These membranes serve as barriers between different compartments in the body to confine certain processes to specific locations. The transport of various molecules and ions through these membranes is of critical importance for the maintenance of a constant internal environment in the body (**homeostasis**).

STOP AND THINK

There can be considerable variation in the permeabilities of body membranes. For instance, compare the permeabilities of the blood-brain barrier and the filtration membrane in the nephron. One is much more permeable than the other. Can you determine why this difference exists? In the following activities, we will examine some of the physiological principles that govern movement through membranes in general and the cell membrane in particular.

Diffusion

Diffusion is the random movement of molecules due to their internal kinetic energy. This continuous movement allows molecules and ions to be distributed uniformly within a closed space (e.g., the plasma or interstitial space). A **net diffusion** of particles results when there is a difference in concentration between two regions of a system; that is, a **concentration gradient** is established. A net diffusion of particles will then take place from the region of higher concentration to the region of lower concentration, and this diffusion will continue until the system reaches dynamic equilibrium (when the concentration everywhere in the system is equal and when molecular movement is equal in all directions). The various factors influencing the rate of diffusion are conveniently grouped together in what is known as **Fick's law of diffusion**:

$$Q = DA \, (C_1 - C_2)$$

Copyright © 2015, 2011, 2008 Pearson Education, Inc.

in which Q is the diffusion rate, $C_1 - C_2$ is the concentration gradient, A is the cross-sectional area, and D is the diffusion coefficient. The diffusion coefficient is a composite constant that takes into account the temperature, the molecular weight of the substance diffusing, and the nature of the substance through which the first substance is diffusing.

 STOP AND THINK

The diffusion constant, D, is directly proportional to the (Velocity)^2 of a molecule, which in turn is directly correlated with temperature. For the experiments in Activity 2.1, can you predict the effect of temperature on the rate of diffusion? If so, what is your prediction? Explain why the kinetic energy of each experimental setup varies.

▶ **ACTIVITY 2.1**

The Effect of Temperature on the Rate of Diffusion

Materials

☐ 1 g crystals of methylene blue

☐ 3 500-ml beakers

☐ Water, available at 5°C, 25°C, and 50°C

1. Place similar amounts of several crystals of methylene blue in three 500-ml beakers containing 400 ml of water that is, respectively, cold (5°C), room temperature (25°C), and hot (50°C).
2. In the Laboratory Report, record the time required for the dye to become evenly dispersed throughout the beakers. ◀

▶ **ACTIVITY 2.2**

Solution Concentration Using Conductivity Probe

The ability of a solution to conduct electricity depends upon the concentration of ions in the solution. The greater the concentration of ions, the more the solution can conduct electricity. Sodium chloride (NaCl) ionizes in water and so is capable of conducting electricity. Based on knowledge of the conductivity of standard concentrations of a specific salt in a solution, we can estimate the concentration of that salt in a solution of unknown concentration if the conductivity is known. In this activity, the conductivity probe will be used to determine the conductivity of four NaCl standard solutions. You

will initially create a graph of conductivity versus NaCl concentration and then use this graph to investigate the effect of concentration gradients on rates of diffusion.

2.2 VERNIER® VERSION

Materials (per group)

☐ 1 computer

☐ 1 conductivity probe

☐ 1 Vernier® LabPro

☐ 200 ml each of 0.1%, 0.5%, 1%, and 3% NaCl solutions

☐ 1 500-ml wash bottle of distilled water

☐ Paper tissues

Materials needed for this exercise will have been set up for you.

1. Set up the conductivity probe, using a ring stand and utility clamp. The conductivity probe is already attached to the interface box and computer.
2. Prepare the computer to monitor conductivity by opening the file in the Experiment 5 folder of *Biology with Computers*. The meter window will display live conductivity readings in units of microsiemens per centimeter (μS/cm).
3. Make sure that the conductivity probe is set to 0–20,000 μS/cm (equivalent to 10,000 mg/L).
4. If necessary, calibrate the conductivity probe, using the conductivity of air (0 μS/cm) and the calibration solution that comes with the probe (1,000 μS/cm).
5. Create a standard curve by using the conductivity of air and determining the conductivity of the four prepared NaCl solutions (0.1%, 0.5%, 1%, 3%) as described below. You can do the tests in any sequence. Obtain and wear goggles for this activity.
6. Place about 50–60 ml of the NaCl solution in a 100-ml beaker.
7. Carefully raise each beaker and its contents up around the conductivity probe until the hole near the probe end is completely submerged in the solution being tested.
8. Briefly stir the beaker contents. When the conductivity reading in the meter window has stabilized, record the value in the table in the Laboratory Report.
9. Before testing the next solution, clean the electrodes by surrounding them with a 250-ml beaker and rinse them with distilled

Copyright © 2015, 2011, 2008 Pearson Education, Inc.

water from a wash bottle. Dry the outside of the probe with a tissue. It is not necessary to dry the inside of the hole near the probe end.

10. Repeat the process for the other NaCl standards and plot a graph of conductivity (μS/cm) versus NaCl concentration on standard graph paper. This will be your standard curve.

2.2 POWERLAB® VERSION

Materials (per group)

☐ 1 computer

☐ 1 PowerLab® data acquisition unit

☐ 1 conductivity pod

☐ 1 conductivity electrode

☐ 200 ml each of 0.1%, 0.5%, 1%, and 3% NaCl solutions

☐ 1 500-ml wash bottle of distilled water

☐ Paper tissues

Materials needed for this exercise will have been set up for you.

1. Locate and launch the Chart software on the computer.
2. Follow the manufacturer's instructions to adjust the sensitivity of the probe, using the conductivity pod window to calibrate the probe.
3. Create a standard curve by using the conductivity of air (= 0 μS/cm) and determining the conductivity of the four prepared NaCl solutions (0.1%, 0.5%, 1%, 3%) as described below. You can do the tests in any sequence. Obtain and wear goggles for this activity.
4. Place about 50–60 ml of the NaCl solution in a 100-ml beaker.
5. Carefully raise each beaker and its contents up around the conductivity probe until the end is completely submerged in the solution being tested.
6. In Chart, click **Start**.
7. Briefly stir the beaker contents. Record for 20 sec.
8. Click **Stop**.
9. Record the conductivity in the table in the Laboratory Report.
10. Rinse the tip of the probe with distilled water.
11. Repeat the process for the other NaCl standards and plot a graph of conductivity (μS/cm) versus NaCl concentration on standard graph paper. This will be your standard curve. ◀

▶ ACTIVITY 2.3

Concentration Gradients and Rate of Diffusion

In this activity you will test the effect of concentration gradients on rates of diffusion using 1%, 5%, and 10% NaCl solutions. Each of these solutions will be placed in a cylinder of sealed dialysis tubing, which, in turn, will be placed in a beaker of distilled water. The diffusion of salt into the distilled water will increase the conductivity of the distilled water.

⬤ STOP AND THINK

Dialysis is a treatment used to remove wastes and excess water from the blood in individuals experiencing the loss of kidney function. It depends upon the physical principles of diffusion of solutes and water across a semipermeable membrane. While there are different methods of dialysis, most depend upon blood flowing on one side of a semipermeable membrane and the dialysis fluid flowing in the opposite direction on the other side. This counter-current flow maximizes the concentration gradient, allowing for the removal of urea and creatinine (both products of metabolic activities in our bodies). In general the concentration of electrolytes like K^+ and Ca^{++} is about the same as that seen in the plasma. These electrolyte concentrations in the dialysis fluid are maintained and constantly renewed so as to maintain homeostasis. HCO_3^- levels in the dialysis fluid are typically higher than in plasma to allow diffusion into the blood. This is done to neutralize the acidosis typically seen in patients with kidney failure. Given that most of the electrolytes in the dialysis fluid are isosmotic with the plasma and given the goals of dialysis, why do "wastes" like urea and creatinine get removed, resulting in filtered plasma?

2.3 VERNIER® VERSION

Materials (per group)

☐ 1 computer

☐ 1 conductivity probe

☐ 1 Vernier® LabPro

☐ 40 cm dialysis tubing, such as Spectra/Por® 12-14 kDa MWCO or Cellu·Sep® Regenerated Cellulose Tubular Membranes [Molecular weight cut-off (MWCO) is a term used to describe the pore size of a membrane. The smaller the MWCO, the tighter the membrane pore size. A membrane with MWCO of 100 can optimally reject molecules with the molecular weight of 100 (i.e. >90%).]

☐ 200 ml each of 1%, 5%, and 10% NaCl solutions

☐ 1 500-ml wash bottle of distilled water

☐ 1 400-ml beaker

☐ 1.5 L distilled water

☐ Paper tissues

Copyright © 2015, 2011, 2008 Pearson Education, Inc.

Materials needed for this exercise will have been set up for you.

1. Prepare a 10-cm piece of dialysis tubing by soaking it in warm water for a few minutes and then rubbing it between your fingers to separate the membranes. Seal one end of the tube by tying it or by using a weighted tubing clamp.
2. Prepare the computer to monitor conductivity by opening the file in the Experiment 5 folder of *Biology with Computers*. The meter window will display live conductivity readings in units of microsiemens (μS/cm).
3. Make sure that the conductivity probe is set to 0–20,000 μS/cm (equivalent to 10,000 mg/L).
4. Transfer about 15 ml of 1% NaCl solution into the tubing, using a pipette or a funnel.
5. Use a nonweighted clamp to seal off the top of the dialysis tube (or tie it) while minimizing the amount of air in the resulting bag.
6. Wash the outside of the bag with tap water. Dry the outside with a paper towel. Check for leaks.
7. Place 300 ml of distilled water in a 400-ml beaker. Immerse the conductivity probe into the water, making sure that the electrodes are submerged. Record the conductivity of distilled water after the reading stabilizes.
8. Place the dialysis bag into the beaker, making sure that it is completely submerged. Maintain the same distance between the probe and the bag when you repeat the experiment using the remaining concentrations of NaCl.
9. Record the conductivity of the solution in the beaker at intervals of 1, 2, 3, 4, and 5 min in the table in the Laboratory Report. Continue to stir the water while data is being collected. After the final reading, remove the dialysis bag from the beaker and discard it.
10. Transfer the conductivity to the standard curve to estimate the NaCl concentration in the beaker at the end of each time interval. Record these values in the table in the Laboratory Report.
11. Determine the change in NaCl concentration in the beaker for each time interval.
12. Using a fresh piece of dialysis tubing, repeat steps 4–11 for the 5% and the 10% NaCl solutions. Before testing the next solution, clean the electrode by surrounding it with a 250-ml beaker and rinse with distilled water from a wash bottle. Dry the outside of the probe with a tissue. It is not necessary to dry the inside of the hole near the probe end.

2.3 POWERLAB® VERSION

Materials (per group)

- ☐ 1 computer
- ☐ 1 conductivity electrode
- ☐ 1 conductivity pod
- ☐ 40 cm dialysis tubing
- ☐ 200 ml each of 1%, 5%, and 10% NaCl solutions
- ☐ 1 500-ml wash bottle of distilled water
- ☐ 1 400-ml beaker
- ☐ 1.5 L distilled water
- ☐ Paper tissues

Materials needed for this exercise will have been set up for you.

1. Locate and launch the Chart software on the computer.
2. Follow the manufacturer's instructions to adjust the sensitivity of the probe, using the conductivity pod window, and to calibrate the probe in microsiemens (μS/cm).
3. Prepare a 10-cm piece of dialysis tubing by soaking it in warm water for a few minutes and then rubbing it between your fingers to separate the membranes. Seal one end of the tube by tying it or by using a weighted tubing clamp.
4. Transfer about 15 ml of 1% NaCl solution into the tubing, using a pipette or a funnel.
5. Use a nonweighted clamp to seal off the top of the dialysis tube (or tie it) while minimizing the amount of air in the resulting bag.
6. Wash the outside of the bag with tap water. Dry the outside with a paper towel. Check for leaks.
7. Place 300 ml of distilled water in a 400-ml beaker. Immerse the conductivity probe into the water, making sure that the electrodes are submerged.
8. In Chart, click **Start**. Record the conductivity of distilled water after the reading stabilizes and then click **Stop**.
9. Place the dialysis bag into the beaker, making sure that it is completely submerged. Maintain the same distance between the probe and the bag when you repeat the experiment, using the remaining concentrations of NaCl.
10. In Chart, click **Start**. Record the conductivity of the solution in the beaker at intervals of 1, 2, 3, 4, and 5 min in the table in the Laboratory Report. Continue to stir the water while data is being collected. After the final reading, click **Stop** and remove the dialysis bag from the beaker and discard it.

Copyright © 2015, 2011, 2008 Pearson Education, Inc.

11. Transfer the conductivity to the standard curve to estimate the NaCl concentration in the beaker at the end of each time interval. Record these values in the table in the Laboratory Report.
12. Determine the change in NaCl concentration in the beaker for each time interval.
13. Using a fresh piece of dialysis tubing, repeat steps 3–12 for the 5% and the 10% NaCl solutions. Before testing the next solution, clean the electrode by surrounding it with a 250-ml beaker and rinse with distilled water from a wash bottle. Dry the outside of the probe with a tissue. It is not necessary to dry the inside of the hole near the probe end. ◄

INQUIRY-BASED ACTIVITY

Appendix C describes the format of a typical Laboratory Report. It is mandatory that you read the Appendix before you start planning your experiment. For both parts, you will utilize the materials and procedures listed in Activity 2.3.

2.3a

Design an experiment in order to study the effects of temperature on diffusion rates. Before you actually proceed, you will need to briefly describe your experiment and make some predictions about the expected results.

2.3b

Here you will design an experiment in order to study the effects of membrane pore size on diffusion rates. In addition to your other materials, you will be provided with dialysis membranes of two pore sizes (6-8 kDa MWCO and 12-14 kDa MWCO). Before you actually proceed, you and your laboratory partners will need to briefly describe your experiment and make some predictions about the expected results.

For each experiment, you need to state what you will measure to demonstrate the differences between your treatments. Use the standard curve created in Activity 2.2 to evaluate the changes in concentration over time. Use the table (or a similar table) in Diffusion: Part 5 of the Laboratory Report to record your data.

Complete this checklist to make sure that you have covered all bases before you start your experiment:

What is the question you seek to answer?

Frame it in the form of a hypothesis.

What is the independent variable or treatment?

How will the independent variable or treatment be measured?

What is the control treatment?

How will you replicate your experiment?

How will you ensure that your subjects are similar enough to not introduce some other independent variability?

Are there any standardized variables?

What is/are the dependent variable(s)?

How will the dependent variable(s) be measured?

What are your predictions?

Construct a table to record your observations easily.

How will you present the data collected graphically?

How will you analyze the data collected?

How will you know if the differences between the treated and the untreated samples are statistically significant?

Your laboratory instructor will describe how your Laboratory Report will be written.

Osmosis

The phenomenon of osmosis occurs whenever a higher concentration of solute is separated from a lower concentration of solute by a membrane that is either semipermeable or selectively permeable. In this situation, water molecules begin moving through the membrane into the region of higher solute concentration. This net movement of water is defined as **osmosis**, and the force of the water movement across the membrane is called the **osmotic pressure**. The amount of osmotic pressure developed depends on the *number* of particles present on either side of the membrane; the more particles there are, the higher the osmotic pressure. Osmotic pressure may be calculated using the following formula:

$$\pi = iRT \left(C_1 - C_2 \right)$$

in which T is absolute temperature in Kelvin; π is the osmotic pressure in atmospheres; i is the number of ions dissociated from each molecule in solution; R is the gas constant, 0.082 L-atm/deg-M; and $C_1 - C_2$ is the concentration gradient in moles per liter.

One mole of an electrolyte can produce more osmotic pressure than can one mole of a nonelectrolyte because the electrolyte will dissociate into two or more ions in solution, each of which acts as an osmotically active particle. To compare the osmotic ability of various solutions, we use the **osmolar** unit of concentration, which was developed for this purpose.

▶ ACTIVITY 2.4
Osmotic Pressure
Materials (per group)

☐ 50 cm dialysis tubing

☐ 2 1 m small bore glass tubing

Copyright © 2015, 2011, 2008 Pearson Education, Inc.

☐ 250 ml 30% glucose solution with 0.25 gm Congo Red crystals

☐ 250 ml 60% glucose solution with 0.25 gm Congo Red crystals

☐ 2 Ring stands and clamps

☐ 2 500 ml beakers

☐ 800 ml distilled water

☐ String

1. To demonstrate the phenomenon of osmosis and osmotic pressure, set up two osmometers.
2. Soak a piece of dialysis tubing, about 15 cm in length, in distilled water for a few minutes until it is pliable.
3. Tie a knot in one end of the tubing and fill a dialysis bag with a 30% sucrose solution (containing a small amount of dye such as Congo Red). Tie the bag to one end of a 3- to 4-ft-long piece of glass tubing with a small bore. This is best done by wrapping heavy cord around the bag and tube several times and tying with a knot.
4. Attach the osmometer to a burette clamp and lower the dialysis bag into a beaker of distilled water (Figure 2.1). If sucrose leaks from the bag, remove the bag from the beaker and tie it more tightly.

5. Mark the height of the fluid in the glass tubing with a marking pencil and record the time. Repeat this procedure with the other osmometer, using a 60% sucrose solution.

In the Laboratory Report

6. Record the number of millimeters the fluid rises for each of the two solutions every 10 min for at least 1 hr. Which has the fastest rate of fluid movement and largest total movement?
7. Plot the millimeters of fluid movement for each 10-min period against the time. How does the rate of fluid movement vary for each of the 10-min periods of time? Why does this change occur? How does this rate of movement compare for the two solutions?
8. Calculate the initial osmotic pressure (in atmospheres) developed by each solution. ◀

▶ **ACTIVITY 2.5**

Osmosis in Plant Cells

Materials (per group)

☐ 1 cork borer

☐ 1 large potato

☐ 1 10-ml graduated cylinder

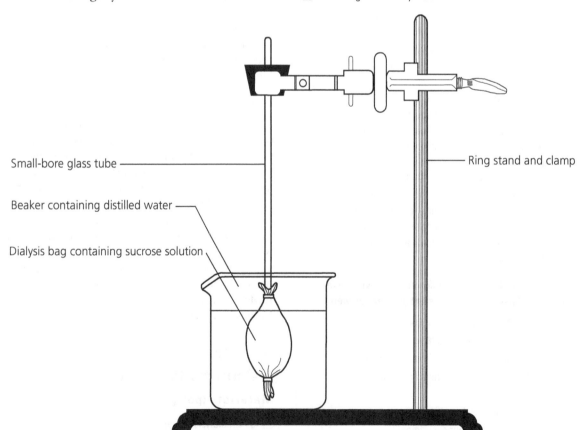

Small-bore glass tube

Beaker containing distilled water

Dialysis bag containing sucrose solution

Ring stand and clamp

FIGURE 2.1 Osmometer setup.

Copyright © 2015, 2011, 2008 Pearson Education, Inc.

Graduated cylinder with known volume of water

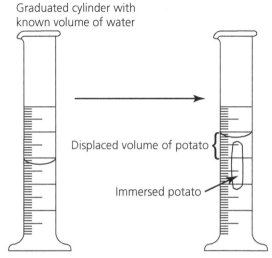

Displaced volume of potato

Immersed potato

FIGURE 2.2   Osmosis by potato.

- ☐ 200 ml each 0.4%, 0.9%, 5%, and 10% NaCl solutions
- ☐ 200 ml distilled water
- ☐ 5 250-ml beakers
- ☐ 1 metric ruler

The cell membranes of plants are similar in their structure to the cell membranes of animals and therefore can also be used to demonstrate various phenomena of membrane transport.

1. Using a cork borer approximately 8–10 mm in diameter, cut five pieces out of a potato, each measuring about 50 mm long.
2. Determine the volume of each piece by immersing it in a known volume of water in a 10-ml graduated cylinder and noting the volume of the rise of water in the cylinder (Figure 2.2).
3. Place one potato piece in each of the following five solutions: distilled water, 0.4% NaCl (sodium chloride), 0.9% NaCl, 5% NaCl, and 10% NaCl. After 2 h, measure the volume of each piece of potato again and express the change as a percent of the original volume.

$$\% \text{ change} = \frac{\text{Change in volume(ml)}}{\text{Original volume(ml)}} \times 100 \blacktriangleleft$$

Tonicity

Within a living cell, all of the molecules, ions, and particles combine to produce a total osmotic pressure. Any solution that contains an equal number of osmotically active particles will produce the same osmotic pressure as that produced by the cellular constituents and is said to be **isotonic** to the cell. Isotonic means "same tension" and implies that the solution will produce no change

in the cell size due to net water movement in or out of the cell. A 0.3-M (300 mM) solution of a **nonelectrolyte**, such as glucose, is isotonic to mammalian cells, as is a 0.9% solution of NaCl (an **electrolyte**). A **hypertonic** solution is one that exerts a greater osmotic pressure than do the cell contents, whereas a **hypotonic** solution produces a lower osmotic pressure than that within the cell.

▶ ACTIVITY 2.6
Tonicity
Materials (per group)

- ☐ 25 ml citrate mammalian blood
- ☐ 50 ml each of 0.2%, 0.4%, 0.6%, 0.9%, 2%, and 5% NaCl solutions
- ☐ 50 ml soap solution
- ☐ 8 10-ml test tubes
- ☐ 5 clean slides and cover slips
- ☐ Wooden toothpicks
- ☐ 1 microscope with 40x objective

1. Macroscopic Observations

1. Arrange a series of labeled test tubes containing 2 ml each of the following solutions: soap solution, distilled water, and NaCl solutions of 0.2%, 0.4%, 0.6%, 0.9%, 2%, and 5%.
2. Add two drops of citrated mammalian blood to each of the test tubes and mix by gently inverting the test tube several times.
3. In the Laboratory Report, record the lysis time (time when the membranes of all red blood cells have been ruptured) for each solution. For the end point of lysis, use the time at which you can first see the lines on a piece of ruled notebook paper held behind the test tube. When the red blood cells are intact, the solution has an opaque, milky appearance; when the cells are lysed, the solution becomes transparent.

2. Microscopic Observations

1. Place a small drop of each of the following solutions on separate, clean microscope slides: 5% NaCl, 0.9% NaCl, 0.4% NaCl, and distilled water.
2. Add a few red blood cells to each drop by dipping a clean toothpick into a drop of blood and then washing it in the drop of solution on the microscope slide. Gently stir. Add a cover slip and examine as rapidly as possible.
3. Compare the cell size and shape in each solution and explain what has happened. Which

Copyright © 2015, 2011, 2008 Pearson Education, Inc.

of the solutions are isotonic, hypertonic, or hypotonic? Figure 2.3 shows possible observations through the microscope. ◀

⬢ STOP AND THINK

The cell membrane is made up primarily of phospholipids, sphingolipids, and cholesterol. A significant amount of protein and carbohydrate is also associated with the membrane. Cholesterol, which is mostly hydrophobic, helps make the membranes impermeable to water. Additionally, the space between membrane molecules is very small. However, some molecules, such as carbon dioxide, oxygen, and lipids, have no problem passing through the cell membrane. What characteristics do these molecules have that make that possible?

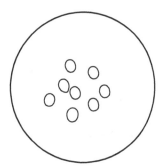

Normal RBCs

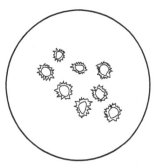

Crenated (shrunk) RBCs

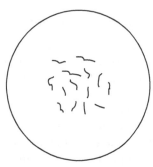

Fragments of Lysed RBCs

FIGURE 2.3   Red blood cells (RBCs) in solutions of various concentrations of sodium chloride.

Cell Permeability

For a substance to enter the cell, it must pass directly through the cell membrane (which is largely lipid) or through the aqueous "pores" in the membrane. The effect of molecular size and lipid solubility on the movement of substances through cell membranes is demonstrated in the following activities.

▶ ACTIVITY 2.7
Molecular Size and Cell Permeability
Materials (per group)

☐ 10 ml each of 0.5 M urea, glycerine, glucose, and sucrose

☐ 10 ml citrated mammalian blood

1. Set up test tubes containing 2–3 ml of each of the following solutions (all 0.5 M): urea (MW 60), glycerine (MW 92), glucose (MW 180), and sucrose (MW 342).
2. Add two drops of citrated mammalian blood to each of the four solutions and mix gently. Determine the time required for lysis of the cells (clearing of the initial murky solution).
3. Observe each solution carefully for the first 3 min after adding the blood. If lysis does not occur within 3 min, go on to the next tube. If you do not observe lysis in a given tube in 30 min, you can assume it will not occur. Because time requirements for different substances vary, start the experiment early in the laboratory period.
4. Determine from the data whether any correlation exists between lysis time and the molecular weights of the substances. In the Laboratory Report, plot hemolysis time against molecular weight. ◀

▶ ACTIVITY 2.8
Lipid Solubility and Cell Permeability
Materials (per group)

☐ Fresh beet slices (small and of approximately the same size)

☐ 250 ml 0.9% NaCl

☐ 6 10-ml test tubes

☐ 10 ml each of 22 M methyl alcohol, 8.5 M ethyl alcohol, 3 M propyl alcohol, 1.1 M isobutyl alcohol, 1.1 M n-butyl alcohol, and 0.38 M amyl (or iso-amyl) alcohol

Copyright © 2015, 2011, 2008 Pearson Education, Inc.

Alcohols can rupture the cell membranes of beets and thereby allow the red pigment **anthrocyanin** to be released from the cells.

1. Cut small slices of fresh beet and *rinse them thoroughly* in 0.9% NaCl to remove any anthrocyanin that may be present on the beet surface.
2. Place one slice of beet into each of six test tubes, each containing one of the alcohols listed in the following table.

In the Laboratory Report

1. Record the time required for the first release of the red anthrocyanin from the beet cells, designated as the **penetration time**.
2. Calculate the **penetration coefficient** for each alcohol by dividing the penetration time in minutes by the concentration of each alcohol.
3. Plot the penetration coefficient against the partition coefficient for each alcohol. The partition coefficient is a measure of a substance's relative affinity for octanol versus water. The higher it is, the more the substance is attracted to octanol and fat-based solvents; the lower it is, the more the substance is attracted to water.

Concentration and Type of Alcohol	Partition Coefficient
22 M methyl alcohol	0.01
8.5 M ethyl alcohol	0.03
3 M propyl alcohol	0.13
1.1 M isobutyl alcohol	0.18
1.1 M n-butyl alcohol	0.58
0.38 M amyl (or iso-amyl) alcohol	2.0

CLINICAL APPLICATION

Transdermal Medications

The skin and its cells are the first line of defense against infective agents such as bacteria and viruses; the skin is a very effective physical barrier. Some drugs, such as scopolamine (for motion sickness), nicotine, estrogen, and nitroglycerine (a vasodilator used to treat heart ailments), are often delivered via medicated adhesive patches that are placed on the skin. How do these patches penetrate the skin's barrier? Transdermal medications have molecules that are both lipophilic and hydrophilic, and they have a small molecular weight that is usually lower than 1,000 daltons. [A dalton is a unit indicating mass on an atomic or molecular scale; 1 Da = 1/12 of the mass of a neutral atom of carbon.] Such medications may even be combined with permeation enhancers, which are substances that promote the passage of molecules through the skin; this facilitates drug absorption into the subcutaneous blood vessels. Transdermal drug delivery has several advantages over oral methods: it maintains even levels of medicine in the bloodstream, it allows for precise dosing by increasing or decreasing the size of the medicated patch, and it enables patients to use a lower dose of the drug than might be available through oral dosing methods.

Copyright © 2015, 2011, 2008 Pearson Education, Inc.

Copyright © 2015, 2008 Pearson Education, Inc.

Renal Physiology

CHAPTER 3 INCLUDES:

 Vernier   1 Vernier® Activity

PowerLab   1 PowerLab® Activity

PhysioEx™ 9.I

For more exercises on Renal Physiology, visit PhysioEx™ (www.physioex.com) and choose Exercise 9: Renal System Physiology.

⚠ **CAUTION!**

Parts of this lab involve experiments with urine. Disposable latex or nitrile gloves must be worn, and the protocol in "Precautions for Handling Blood" must be followed. (See "General Safety in the Laboratory" inside the front cover of this text.)

OBJECTIVES

After completing this exercise, you should be able to

1. Demonstrate the function of the kidney when presented with an excess salt or water load.
2. Quantify variations in urine content when presented with an excess salt or water load.
3. Understand the role of the kidney in the maintenance of homeostasis.
4. Use urinanalysis strips to examine and understand disorders of kidney function.

Kidney Regulation of Osmolarity

The organs responsible for the main function of the urinary system are the kidneys. These paired organs, located in the abdominal cavity, are responsible for filtering the blood and then modifying the resultant filtrate to form urine. This process of filtration occurs continuously, with the plasma passing through the kidneys about 60 times a day. The resulting 180 liters of filtrate formed is ultimately converted to approximately 1.5 liters of urine. This is due to two other kidney functions—reabsorption and secretion— which together bring about the formation of a low-volume and (usually) highly concentrated

urine (to 1,200 mOsm). All of these functions take place in the functional unit of the kidney, the nephron (Figure 3.1). Each kidney has about a million of these structures.

One of the kidneys' main functions is to regulate the osmolarity of the body fluids at around 300 mOsm/L. Regulation of osmolarity will be demonstrated in the following activity by presenting the kidney with an excess water or salt load and recording its response as reflected in the concentration and volume of urine produced.

⬤ **STOP AND THINK**

Because the formation of the filtrate depends on the hydrostatic pressure and the colloid osmotic pressure of blood and the corresponding hydrostatic pressure and colloid osmotic pressure of the filtrate in Bowman's capsule, how would drinking the fluids in Activity 3.1 affect urine volume and concentration?

▶ **ACTIVITY 3.1**

Kidney Regulation of Osmolarity

Materials (per group of 6 or 8 students)

☐ 12 urine collection cups

☐ 4 L drinking water

☐ 3 or 4 pints Gatorade® or Powerade®

☐ 1 urinometer

☐ 20 ml of 20% potassium chromate

☐ 25 ml of 2.9% silver nitrate in amber bottle

☐ 1 500-ml measuring cylinder

☐ 2 sheets graph paper per student

Copyright © 2015, 2011, 2008 Pearson Education, Inc.

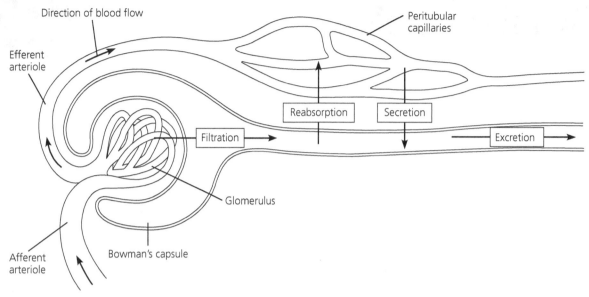

FIGURE 3.1   Structure and function of the nephron.

Source: Stanfield, Cindy L.; Germann, William J., *Principles of Human Physiology*, 3rd Ed., © 2008, pg. 517. Reprinted and electronically reproduced by permission of Pearson Education, Inc., Upper Saddle River, New Jersey.

1. Limit your fluid intake on the day of the experiment. Empty your bladder 1 or 2 hr before the laboratory begins and record the exact time. Do not save this urine sample.
2. On entering the laboratory, take a urine collection bottle to the restroom and void into the bottle, emptying your bladder completely. Record the exact time. This will be designated the control urine.
3. Return to the laboratory and immediately drink the solution assigned to you as quickly as possible. The class will be divided into two groups as follows:

   > Group 1 drinks 800 ml of water.
   > Group 2 drinks 800 ml of an isotonic sports drink such as Gatorade® or Powerade®.
   > [A hypothetical group 3 drinks 80 ml of water plus 7 g of NaCl (coated salt tablets).]

4. Every 30 min after drinking the solution, empty your bladder into a clean collection bottle. If you are unable to void, retain the urine in the bladder until the next 30-min collection time.
5. Analyze the urine from each collection for the following:

   a. Volume. Measure the total volume with a graduated cylinder and express it as milliliters per minute excreted.
   b. Specific gravity. Pour some of the urine sample into a urinometer cylinder and measure the specific gravity as described in the section of Activity 3.2 on specific gravity.
   c. Chloride concentration. Place 10 drops of urine into a test tube (use a standard medicine dropper, 20 drops/ml).

   Add 1 drop of 20% potassium chromate. Add 2.9% silver nitrate solution drop by drop, shaking the mixture continuously while the silver nitrate is being added. Count the number of drops of silver nitrate required to change the bright yellow solution to a brown color. Each drop of 2.9% silver nitrate added to produce the brown color represents 1 g/l (1 mg/ml) of NaCl in the urine. The silver nitrate should be made fresh daily.
   Calculate the total grams of NaCl in the urine collected at each 30-min period and the NaCl concentration in milligrams per milliliter.

In the Laboratory Report

6. Record the data from all students on the data sheet, calculate average values for each experimental group, and graph the results. Make *three separate plots* of milliliters of urine per minute, specific gravity, and chloride concentration (mg/ml), each plotted against time. Use different colors on each plot to represent the average results of groups 1, 2, and 3. Be able to explain how the results illustrate the kidneys' processing of the water or salt loads.

The sodium chloride concentration of urine can also be estimated using the conductivity probes and the procedures described in Chapter 2.

Copyright © 2015, 2011, 2008 Pearson Education, Inc.

3.1 VERNIER® VERSION

Materials (per group)

- [ ] Urine collection cups
- [ ] 1 computer
- [ ] 1 conductivity probe
- [ ] 1 Vernier® LabPro
- [ ] 1 500-ml wash bottle of distilled water
- [ ] 1 400-ml beaker
- [ ] 1.5 L distilled water
- [ ] Paper tissues

Materials needed for this activity will have been set up for you.

1. Prepare the computer to monitor conductivity by opening the file in the Experiment 5 folder of *Biology with Computers*. The meter window will display live conductivity readings in units of microsiemens per centimeter (μS/cm).
2. Make sure that the conductivity probe is set to 0–2,000 μS/cm (equivalent to 1,000 mg/l). This setting may need to be adjusted to 0–20,000 μS/cm (equivalent to 10,000 mg/l) if necessary.
3. Place about 50–60 ml of urine solution in a 100-ml beaker.
4. Carefully raise the beaker and its contents up around the conductivity probe until the hole near the probe end is completely submerged in the urine being tested.
5. Briefly swirl the beaker contents. Once the conductivity reading in the meter window has stabilized, record the value in the table in the Laboratory Report.
6. Before testing the next solution, clean the electrodes by surrounding them with a 250-ml beaker and rinsing them thoroughly with distilled water from a wash bottle. Dry the outside of the probe with a tissue. It is not necessary to dry the inside of the hole near the probe end.
7. Repeat the process for subsequent urine samples, recording the conductivity for each sample in the Laboratory Report.
8. Use the standard curve to estimate the concentration of NaCl for each time period. From this you can calculate the total grams of NaCl in the urine collected at each 30-min period and the NaCl concentration in milligrams per milliliter.

3.1 POWERLAB® VERSION

Materials (per group)

- [ ] Urine collection cups
- [ ] 1 computer
- [ ] 1 PowerLab® data acquisition unit
- [ ] 1 conductivity electrode
- [ ] 1 conductivity pod
- [ ] 1 500-ml wash bottle of distilled water
- [ ] 1 400-ml beaker
- [ ] 1.5 L distilled water
- [ ] Paper tissues

Materials needed for this activity will have been set up for you.

1. Place about 20–30 ml of urine solution in a 100-ml beaker. Carefully raise the beaker and its contents up around the conductivity probe until the probe end is completely submerged in the urine being tested.
2. Launch Chart. Follow the manufacturer's instructions to adjust the sensitivity of the probe and to calibrate the probe.
3. In Chart, click **Start**.
4. Briefly stir the beaker contents. Record for 20 seconds.
5. Click **Stop**.
6. Remove the electrode from the urine; rinse the tip in your waste beaker with distilled water. Dry the outside of the tip with a tissue.
7. Record the conductivity in the table in the Laboratory Report.
8. Repeat the process for subsequent urine samples, recording the conductivity for each sample. Use a clean beaker for each urine sample.
9. Use the standard curve to estimate the concentration of NaCl for each time period. From this you can calculate the total grams of NaCl in the urine collected at each 30-min period and the NaCl concentration in milligrams per milliliter. ◄

CLINICAL APPLICATION

Chronic Kidney Disease

Chronic kidney disease (CKD) is the gradual loss of kidney function over months or years. The underlying cause for this disease may vary, although diabetes and hypertension are often the reasons. Gradual loss in the numbers of functional nephrons will lead to symptoms that include uremia (accumulation of urea in the blood), edema (retention of fluid in the tissues), and hyperkalemia (increased potassium in the blood). Although increases in blood creatinine levels due to decreases in glomerular filtration rates typically serve as the diagnostic test for CKD, early stages of the disease may be detected by urinalysis, which will show the presence of red blood cells (RBCs) in the urine for at least three months. Recent guidelines classify CKD into five stages, with stage 1 being the mildest form and stage 5 (or end-stage) being the most severe form of the disease. Although early-stage CKD can be treated medically, stage 5 CKD requires renal replacement therapy in the form of dialysis or kidney transplant.

Copyright © 2015, 2011, 2008 Pearson Education, Inc.

INQUIRY-BASED ACTIVITY

Appendix C describes the format of a typical Laboratory Report. It is mandatory that you read the Appendix before you start planning your experiment.

The experiments described earlier are meant to qualitatively demonstrate the function of the kidney. If one were to use the scientific method to repeat this experiment in order to make it more quantitative, it is clear that you would need to standardize the procedures.

Because the excretory system's function is closely aligned with that of the circulatory system, it should be clear that the volume of liquid consumed could affect the concentration of fluid circulating in the body and, therefore, the osmolarity of blood. Accordingly, it is important to standardize the dilution effect. Students in the laboratory clearly vary in terms of gender and body size; their blood volume would also vary. The Nadler formula is used to estimate blood volume in males and females based upon their height and weight. Use it to standardize the amount of liquid each individual should consume to reduce the variability in the dilution of his or her intracellular and extracellular fluid compartments.

Blood Volume Formula

$$\text{Males} = 0.3669 \times (\text{Ht in M})^3 + 0.03219 \times (\text{Wt in kgs}) + 0.6041$$

$$\text{Females} = 0.3561 \times (\text{Ht in M})^3 + 0.03308 \times (\text{Wt in kgs}) + 0.1833$$

Use the final table in the Kidney Regulation of Osmolarity section of the Laboratory Report to calculate the volume of liquid to be consumed by each student based upon his or her height, weight, and gender. It is suggested that you consume 15 ml of liquid per 100 ml of blood volume.

Complete this checklist to make sure that you have covered all bases before you start your experiment:

What is the question you seek to answer?

Frame it in the form of a hypothesis.

What is the independent variable or treatment?

How will the independent variable or treatment be measured?

What is the control treatment?

How will you replicate your experiment?

How will you ensure that your subjects are similar enough to not introduce some other independent variability?

Are there any standardized variables?

What is/are the dependent variable(s)?

How will the dependent variable(s) be measured?

What are your predictions?

Construct a table to record your observations easily.

How will you present the data collected graphically?

How will you analyze the data collected?

How will you know if the differences between the treated and the untreated samples are statistically significant?

Your laboratory instructor will describe how your Laboratory Report will be written.

Urinalysis

The kidneys are the chief regulators of the internal environment of the body. They do this by regulating the concentration of ions, water, and pH in the various body fluids. In addition, they provide for the elimination of the waste products of metabolism. Each of the million nephrons in the kidneys contains two main structures, the **glomerulus** and the **renal tubule**. As blood passes through the kidneys, it is first filtered through the glomerulus (120 ml/min), and the filtrate then passes into the renal tubule.

The tubular filtrate is similar to blood plasma in composition except that large molecules (molecular weights greater than 70,000) are excluded (e.g., plasma proteins). As this filtrate passes along the proximal and distal tubules, most of the water is reabsorbed, and many essential substances are actively or passively reabsorbed into the bloodstream. Toxic by-products of metabolism and substances in excess are retained in the filtrate or are secreted into the filtrate and finally are excreted in the urine (1 ml formed/min). Thus, the final composition of the urine is quite different from that of the glomerular filtrate and reflects the integrity of kidney function and changes in blood composition.

An analysis of urine can yield valuable information about the health of the kidney and of the body in general. Various diseases are characterized by abnormal metabolism, which causes abnormal by-products of metabolism to appear in the urine. For example, in individuals with phenylketonuria (PKU), phenylpyruvic acid appears in the urine as a result of an autosomal recessive genetic disorder characterized by a deficiency in the enzyme, hepatic phenylalanine hydroxylase (PAH). This enzyme is necessary to metabolize the amino acid phenylalanine into the amino acid tyrosine. When PAH is deficient, phenylalanine accumulates and is converted into phenylpyruvate. Left untreated, this disease results in mental retardation. In diabetes mellitus, deficient production of insulin (type 1 diabetes) by the pancreas or insulin insensitivity (type 2 diabetes) results in the appearance of glucose in the urine (glycosuria). The volume of urine produced and its specific gravity give information about the state of hydration or dehydration of the body.

Copyright © 2015, 2011, 2008 Pearson Education, Inc.

In the following activity, you will analyze your own urine or a provided sample for some of its clinically important constituents. A sample of urine containing abnormal quantities of these constituents will also be examined to allow comparison with your urine.

CLINICAL APPLICATION

Diabetes Insipidus and Diabetes Mellitus

Diabetes insipidus and diabetes mellitus are two conditions that, while having very different causes, share the symptom of causing polyuria. Diabetes insipidus is caused either by a decreased production of ADH (anti-diuretic hormone) from the pituitary or a reduced response to ADH by the kidney. Both of these conditions will lead to reduced water reabsorption in the distal convoluted tube and the collecting duct of the nephron. The urine will not usually contain sugar. Diabetes mellitus, however, causes polyuria due to osmotic diuresis caused by high blood sugar leaking into urine. This increased sugar osmotically draws water into the ultrafiltrate, causing the production of high-volume urine with sugar. Both conditions will lead to significant thirst caused by excessive water loss through polyuria.

▶ ACTIVITY 3.2 URINALYSIS
Materials

☐ Urine collection cups

☐ 1 urinometer

☐ 1 thermometer

☐ Labstix reagent strips (or similar product)

☐ pHydrion strips

☐ Clinitest tablets

Take a urinalysis bottle to the restroom and collect a 15- to 25-ml sample of your urine or use the sample of urine collected for Activity 3.1.

1. Specific Gravity

1. Fill a urinometer cylinder to about 1 in. from the top with the collected urine.
2. Holding the urinometer flat by its stem, slowly insert it into the cylinder. Do not wet the stem above the water line, or an inaccurate reading will result.
3. Give the float a slight swirl and read the specific gravity from the graduated marks on the stem as it comes to rest. Do not accept a reading if the float is against the side of the cylinder.

4. Remove the float, rinse it in tap water, and dry it. Measure the temperature of the urine specimen immediately.
5. Return the urine specimen to the urinalysis bottle and rinse and dry both the urinometer cylinder and the thermometer.
6. The urinometer is calibrated to give a correct reading only if the urine is at 15°C. If your urine is at a different temperature, you will need to correct the specific gravity by adding 0.001 for each 3°C above or by subtracting 0.001 for each 3°C below the calibration temperature (15°C). **Record your results in the Laboratory Report**.

The normal range of urine specific gravity is 1.0015 to 1.035. Readings above or below these limits may indicate a pathological condition. For example, a low reading is found in chronic nephritis.

2. Urinalysis with Labstix
Advances in urinalysis techniques have made it possible to perform, in only a few seconds, tests that previously took hours. The Labstix test is a combined test of urinary pH, protein, glucose, ketones, and occult blood. Abnormally low pH, along with a high level of glucose and ketones, indicates diabetes mellitus. Alkaline urine is found in many conditions, for example, in cystitis, in which urine decomposes in the bladder with the production of ammonia. Urinary pH usually is slightly acidic (around pH 6), but the pH may be lowered by a diet rich in proteins or citrus fruits, so pH alone is not very informative. Protein and occult blood in the urine are much more definite, indicating nephritis, a disease in which the glomeruli are damaged and plasma proteins and erythrocytes leak into the kidney tubules.

1. Obtain a Labstix reagent strip and bottle with the color standards. Examine the strip carefully before making the test so you will know which portions to read first.
2. When ready, dip the reagent portions into the well-mixed urine specimen, wetting all five reagents completely. Wipe the excess urine off on the lip of the urinalysis bottle.
3. Follow the instructions on the bottle containing the test strips. If the urine glucose or pH is found to be beyond the normal range, make a more accurate analysis for glucose by using the Clinitest tablets and for pH by using the pHydrion paper.
4. Obtain a sample of the abnormal urine and test it with the Labstix strip so that you can compare your urine results with some non-normal results. ◀

Copyright © 2015, 2011, 2008 Pearson Education, Inc.

Copyright © 2014 Pearson Education, Inc.

Neuroanatomy

OBJECTIVES

After completing this exercise, you should be able to

1. Examine the organization of the nervous system.
2. Understand the structure and function of the spinal cord and spinal nerves.
3. Explain the innervation patterns and functions of the cranial nerves.
4. Dissect and understand the organization and the functions of the brain.

Organization of the Nervous System

The nervous system is classically organized into the central and peripheral nervous systems as outlined here:

Central nervous system (CNS)
- Brain
- Spinal cord

Peripheral nervous system (PNS)
- Sensory (afferent)
- Motor (efferent)
  - Somatic nervous system
  - Autonomic nervous system (ANS)
    - Sympathetic system
    - Parasympathetic system

Neurons that lie completely within the brain or spinal cord are in the CNS, whereas neurons that lie partially or wholly outside the CNS are considered peripheral neurons. Cell bodies and synapses are often grouped together as functional integrative centers called **nuclei** in the CNS or **ganglia** in the PNS. Sensory neurons carry nerve impulses directed toward (**afferent**) the CNS, whereas motor neurons conduct impulses away from (**efferent**) the CNS to control the various organs of the body. **Somatic** neurons innervate skeletal muscles to produce voluntary body movements. **Autonomic** neurons provide involuntary control of the internal organs that regulate homeostasis of the internal environment. The individual peripheral neurons that relay information to and from the integrating centers of the CNS are grouped into bundles called the **cranial nerves** and the **spinal nerves**.

Spinal Nerves and Spinal Cord

Materials

1 spinal cord and spinal nerves model
1 model of a spinal cord cross-section

The 31 pairs of spinal nerves are all mixed nerves (i.e., they contain both sensory and motor neurons) and are grouped from the top of the body down, as shown in Figure 4.1.

If the spinal cord is cut in cross-section anywhere along its length, a similar anatomical pattern is seen, as diagrammed in Figure 4.2.

Note that near the spinal cord, the spinal nerve branches into the **dorsal root**, which contains only sensory neurons, and the **ventral root**, which contains the motor neurons. In rare instances, the sensory neuron synapses directly with a motor neuron, thus producing a **monosynaptic** reflex arc. More often, however, sensory neurons synapse with **association neurons** (**interneurons**), which then synapse with the motor neurons. This produces a more complex, **multisynaptic** (polysynaptic) spinal reflex. The **gray matter** of the spinal cord (and the brain) is a region containing mainly nerve cell bodies and synapses; the **white matter** is made up of bundles of myelinated axons, which give a white color to the area.

Sensory neurons that enter the spinal cord synapse not only with motor neurons to produce spinal reflexes but also with association neurons, which extend up the cord to provide the brain with sensory information. Figure 4.3 is a simple schematic of the synapses involved in one set of ascending neurons. Note that three neurons are involved and that nerve impulses cross over from one side of the body to the other before reaching the brain. A simplified view of one set of descending neurons is also shown in Figure 4.3. Again, the motor pathways cross over before synapsing with the spinal motor neurons.

Copyright © 2015, 2011, 2008 Pearson Education, Inc.

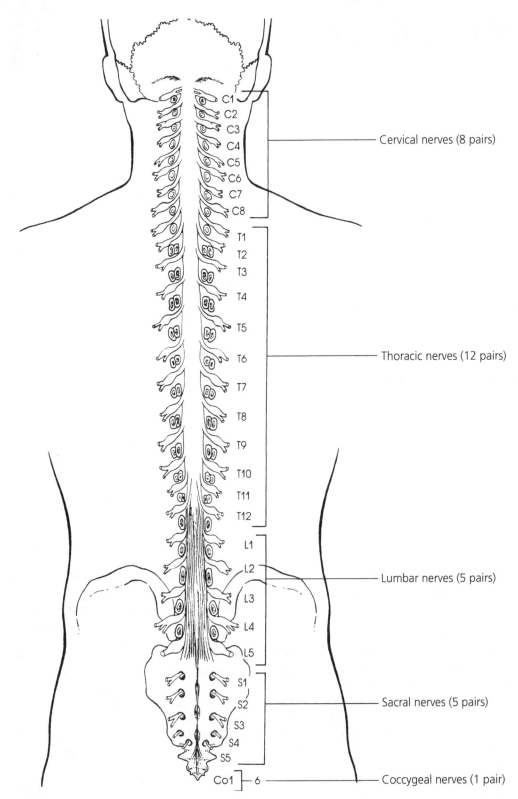

C1
C2
C3
C4
C5
C6
C7
C8

— Cervical nerves (8 pairs)

T1
T2
T3
T4
T5
T6
T7
T8
T9
T10
T11
T12

— Thoracic nerves (12 pairs)

L1
L2
L3
L4
L5

— Lumbar nerves (5 pairs)

S1
S2
S3
S4
S5

— Sacral nerves (5 pairs)

Co1 — 6

— Coccygeal nerves (1 pair)

FIGURE 4.1 Spinal nerves.

The ascending and descending axons in the white matter of the spinal cord are grouped into bundles called **tracts** or **fasciculi**. Most of these tracts are named in a rather simple manner: the prefix indicates where the neuron originates, and the suffix tells where it terminates. Some of the major spinal cord tracts are shown in Figure 4.4.

Copyright © 2015, 2011, 2008 Pearson Education, Inc.

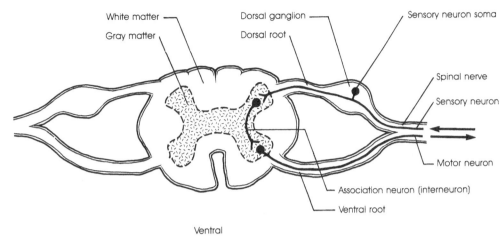

FIGURE 4.2   Spinal cord cross-section.

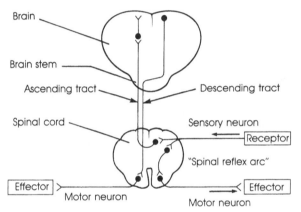

FIGURE 4.3   Generalized scheme of ascending and descending nerve tracts.

● STOP AND THINK

Based upon the anatomy of the spinal cord and the pattern of spinal nerves exiting and entering the spinal cord, would you expect the complexity (in terms of spinal pathways) to increase or decrease as you move from the cervical to the lumbar region? Would a serious spinal cord injury in the cervical or sacral injury produce the greatest harm?

Use the spinal cord model and other figures from your text to become familiar with the anatomy of the spinal cord and the spinal nerves. Use your textbook to learn about the fasciculi and tracts. **In the Laboratory Report, enter the origin and termination points for all the tracts listed.**

Cranial Nerves

Materials (per group)

1 model of a human brain

The 12 pairs of cranial nerves originate from various structures on the ventral surface of the brain and are numbered anterior to posterior, as follows:

I. Olfactory	IV. Trochlear
II. Optic	V. Trigeminal
III. Oculomotor	VI. Abducens

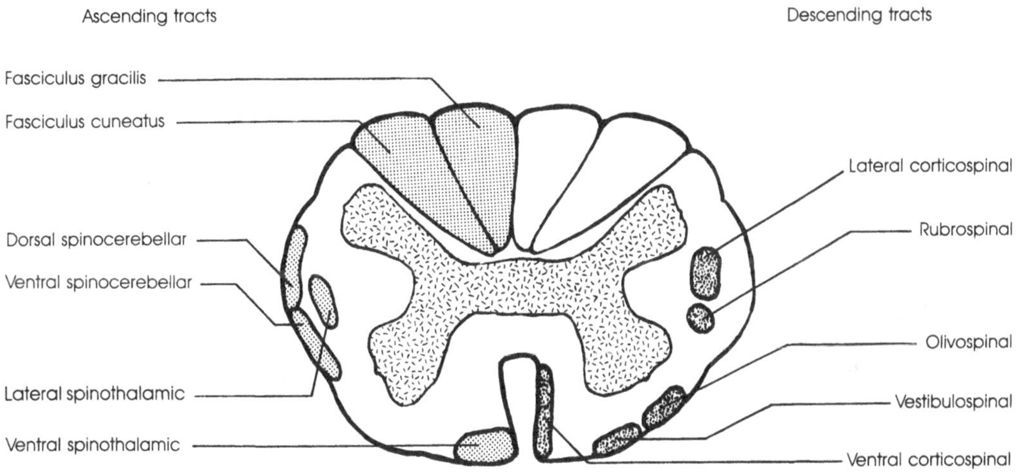

FIGURE 4.4   Major spinal cord tracts.

Copyright © 2015, 2011, 2008 Pearson Education, Inc.

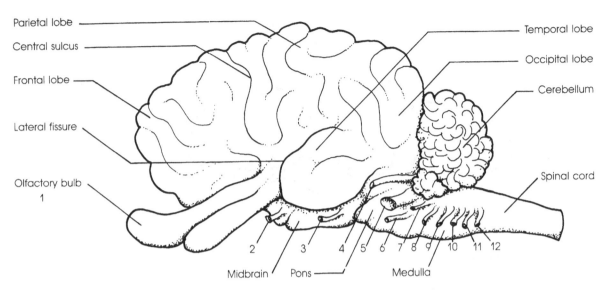

FIGURE 4.5   Sheep brain (lateral view). Numbers correspond to the 12 pairs of cranial nerves.

VII. Facial   X. Vagus
VIII. Vestibulocochlear   XI. Accessory
IX. Glossopharyngeal   XII. Hypoglossal

1. Examine the sheep brain and try to identify as many of these nerves as possible (see Figure 4.5). Some of the nerves may have been lost or damaged when the brain was removed from the skull. Several of the cranial nerves are mixed nerves; some are strictly sensory or motor in their functions.

2. Using your text and other resources, identify the functions of each cranial nerve and its classification (mixed, sensory, or motor). **Enter this information in the table in the Laboratory Report.** To help you remember the cranial nerves in their proper order, you might use this old mnemonic rhyme: "On Old Olympus' Towering Top A Fat Vicious Giant Vaults And Hops." The first letter of each word is the first letter of the corresponding cranial nerve. There are other versions of this mnemonic that, while more entertaining, are not appropriate for a scientific publication.

▶ ACTIVITY 4.1
Cranial Nerve Tests

Materials (per group)

☐ Small vials, each with cotton balls with drops of vanilla, lavender, or other chemicals with strong fragrances

☐ Long-handled cotton swabs

⬤ STOP AND THINK

Many medical examinations will involve your physician asking you to hold your head still while following a moving light or pencil with your eyes. If you pay attention, you will notice that the movements are not random but follow a consistent pattern. Additionally, your physician may darken and brighten the light in the room while examining your pupils. It may surprise you to learn that the examination involves testing the function of your cranial nerves. Which cranial nerves are being tested?

Complete the activities in the cranial nerve tests table in the Laboratory Report and record your observations. ◀

External Structures and Landmarks of the Brain

During embryonic development, the brain changes from a simple neural tube to a complex mass with several prominent enlargements. Examine the brain and identify these structures: spinal cord, medulla oblongata, pons, cerebellum, cerebrum, pituitary, and olfactory bulbs. In higher animals, such as sheep, monkeys, and humans, the cerebrum has grown so large that its surface has developed **fissures** (deep folds) and **sulci** (small folds), which increase its surface area and produce some key landmarks and lobes of the cerebrum. The deepest groove is the **longitudinal fissure**, which separates the left and right cerebral hemispheres. The **lateral fissure of Sylvius** separates the **frontal lobe** from the **temporal lobe**. The **central sulcus**, which separates the frontal lobe

Copyright © 2015, 2011, 2008 Pearson Education, Inc.

from the **parietal lobe**, is less obvious in the sheep brain than in the human brain. The **occipital lobe**, at the rear area of the cortex, is not separated from the parietal and temporal lobes by any major sulci or fissures. Study the external features of the sheep brain until you are familiar with the key structures and landmarks.

CLINICAL APPLICATION

Cardiovascular Accidents (CVAs) or Stroke

When blood supply to any part of the brain is interrupted or severely reduced, nutrients and oxygen are no longer delivered to the cells in that area. Cell death will typically occur in a matter of minutes. Because the areas that are damaged are responsible for regulating or controlling one or more bodily functions, the symptoms of a stroke vary. For example, a victim might experience trouble walking, talking, seeing, or maintaining balance. It is important to understand that the reason for these symptoms is due to neuronal death rather than to problems with the organs involved in those functions.

► ACTIVITY 4.2
Sectioning of the Brain

Materials (per group)

☐ 1 sheep brain
☐ 1 scalpel
☐ 1 forceps

1. Midsagittal Section

1. Using a sharp scalpel, section the sheep brain into right and left halves by cutting through the longitudinal fissure, the corpus callosum, and the entire brain stem (Figure 4.6). To make a clean cut, pull the knife toward you with a continuous movement, anterior to posterior. The midsagittal section allows you to view the following major areas of the brain:

Forebrain   Telencephalon: cerebral hemispheres

Diencephalon: thalamus, hypothalamus

Midbrain   Mesencephalon

Hindbrain   Metencephalon: pons, cerebellum
Myelencephalon: medulla

2. Note the relationship of the third and fourth ventricles to these brain structures. The ventricles of the brain contain the cerebrospinal fluid (CSF), which transports nutrients to the neurons of the CNS. The first and second lateral ventricles in each cerebral hemisphere are not visible in this section but will show in the cross-sections.

2. Cross-Sections

1. Make several cross-sections of the left or right hemisphere from the front of the brain to the back. Do not be concerned with detailed structures, but try to develop in your mind a three-dimensional view of the major structures you have examined previously.

2. **Use your dissections and your textbook to answer the questions in the Laboratory Report.** ◄

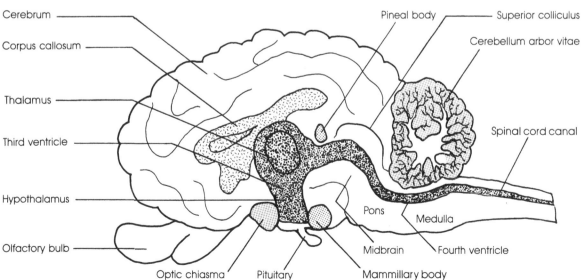

FIGURE 4.6   Sheep brain (midsagittal section).

Copyright © 2015, 2011, 2008 Pearson Education, Inc.

Neuroanatomy

Name _____

Date _____ Section _____

Score/Grade_____

Spinal Nerves and Spinal Cord

Give the origin and termination of each of the following spinal nerve tracts.

Tract	Origin	Termination
Spinothalamic	_____	_____
Rubrospinal	_____	_____
Corticospinal	_____	_____
Spinocerebellar	_____	_____
Fasciculus cuneatus	_____	_____
Olivospinal	_____	_____

Cranial Nerves

1. Give the name, type, and function of each cranial nerve.

Number	Name	Type	Function
I			
II			
III			
IV			
V			
VI			
VII			
VIII			
IX			
X			
XI			
XII			

Copyright © 2015, 2011, 2008 Pearson Education, Inc.

2. Test cranial nerve function.

Cranial Nerve	Test	Yes	No
I	Can you distinguish between the odors and identify them?		
II	Can you count the number of words in this sentence?		
III	Keeping your head still, can you follow the movement of a pencil controlled by your partner up and down and then with each eye (separately) medially? Does your pupil dilate or constrict when you move a pencil closer to or farther away from your eye?		
IV	Can each eye roll down and look to the side?		
V-I	Stroke the upper eyelid, forehead, eyebrow, and upper nose area of your partner. Does he or she feel them equally on both sides of the face?		
V-II	Stroke the lower eyelid, upper gums, lower nose, and upper lip of your partner. Does he or she feel them equally on both sides of the face?		
V-III	Stroke the lower gums, teeth, and lip. Are sensations felt equally on both sides of the face? Close your mouth and clench your teeth. Have your partner place one hand under your chin. Can you open your mouth against this opposing force?		
VI	Can each eye look to the side?		
VII	Can you smile, show your teeth, puff your cheeks, wrinkle your forehead, raise one or both eyebrows?		
VIII	Can you tell whether your partner is closer to your left or right when he or she makes a sound from somewhere behind you?		
IX and X	Have your partner use a moist, long cotton swab to touch the back of your oral cavity, approximately 1 cm behind your hard palate. Do you have a gag reflex? Can you speak normally?		
XI	Can you turn your head to the side and shrug your shoulders against a resistance?		
XII	Does your tongue deviate to one side when you stick it out of your mouth?		

External Structures and Landmarks of the Brain

1. Briefly describe the location in the cerebrum of the following areas.

Primary motor area _____

Primary sensory area _____

Copyright © 2015, 2011, 2008 Pearson Education, Inc.

Primary hearing area _____

Broca's speech area _____

Primary visual area _____

Premotor area _____

2. What is the overall function of the following structures?

Cerebellum _____

Corpus callosum _____

Thalamus _____

Hypothalamus _____

Pituitary (hypophysis) _____

Optic chiasma _____

Pineal body _____

Medulla _____

Hippocampus _____

Basal ganglia _____

APPLY WHAT YOU KNOW

1. What would happen to the area supplied by a spinal nerve if the dorsal root were cut?

_____

If the ventral root were cut? _____

2. If someone wants to compliment you on how smart you are, he or she might say, "You sure have a lot of gray matter." Is this statement truly a compliment? Explain why or why not.

_____

_____

_____

3. Two patients had strokes that caused aphasia, a condition that interferes with a person's ability to communicate. Their symptoms were rather different. One patient had severe problems understanding simple instructions, but he could speak. The other patient could not speak, but could understand verbal instructions with no problem. Identify the type of aphasia in each patient and the brain areas involved.

_____

_____

_____

_____

_____

_____

Copyright © 2015, 2011, 2008 Pearson Education, Inc.

Membrane Action Potentials

5

CHAPTER 5 INCLUDES:

PowerLab   1 PowerLab® Activity

PhysioEx 9.1   For more exercises on Membrane Action Potentials, visit PhysioEx™ (www.physioex.com) and choose Exercise 3: Neurophysiology and Nerve Impulses.

OBJECTIVES

After completing this exercise, you should be able to

1. Describe and understand the physiology of the resting membrane and action potentials.
2. Understand the use of an oscilloscope in the measurement of potential changes involved in an action potential.
3. Record the action potential and demonstrate the physiological properties of the frog sciatic nerve.

Resting and Action Potentials

Various body activities, such as secretion, muscle contraction, mental activity, cardiac function, and sensory perception, involve electrical changes across membranes. These electrical changes are produced by movement of ions, which produces the **resting** and the **action potentials**.

Resting Membrane Potential

The resting membrane exists due to the uneven distribution of positively and negatively charged ions across the cell membrane. The outside surface of the cell membrane has a larger net concentration of positive ions and is normally positive (+). The inside surface of the membrane has a greater net concentration of negative ions and is therefore negative (−) with respect to the outside of the membrane. The average mammalian cell has a resting potential of −85 millivolts (mV) across its membrane. The intracellular fluid will contain approximately 150 mM K^+, 15 mM Na^+, 10 mM Cl^-, and 0.0001 mM Ca^+. The extracellular fluid contains approximately 5 mM K^+, 145 mM Na^+, 108 mM Cl^-, and 1 mM Ca^+. While it is often assumed that no ionic movement occurs when the axon is at rest, it is significant to note the presence of leak channels that are always open and, if not compensated for by the activity of the Na^+/K^+ pump, which maintains the resting membrane potential, would lead to the absence of a resting membrane potential.

Action Potential

When a membrane is stimulated, its permeability to various ions is altered. This results in a series of local changes called the **action potential** (Figure 5.1). Initially, there is a rapid diffusion of Na^+ into the cell due to the rapid opening of sodium channels. This causes the membrane potential to become more positive (+30 mV) locally. This change of polarity is termed **depolarization**. Although K^+ gates also open due to the depolarization induced by the stimulus, they do so slowly, resulting in a subsequent **repolarization** (to −70 mV) as K^+ diffuses out of the cell. This depolarization and repolarization does not remain localized but spreads rapidly (is propagated) down the membrane in both directions. This propagation of the action potential is called the **impulse**. A membrane is capable of transmitting many thousands of impulses (action potentials) before it becomes fatigued.

Stimulation of Tissues

In many experiments, it is necessary to stimulate living tissue in order to record a particular response. Although tissue may be excited (depolarized) by various types of stimuli (e.g., chemical, thermal, mechanical), the method of choice is electrical stimulation. This method is advantageous because it is more closely related to the electrical phenomena occurring in tissues (resting and action potentials) and because it is easy to apply and regulate. The electronic stimulator provides quantitative control

Copyright © 2015, 2011, 2008 Pearson Education, Inc.

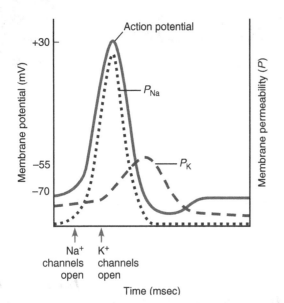

FIGURE 5.1 Permeability changes for Na$^+$ and K$^+$ during an action potential.

Source: Stanfield, Cindy L.; Germann, William J., *Principles of Human Physiology*, 3rd Ed., © 2008, pg. 182. Reprinted and electronically reproduced by permission of Pearson Education, Inc., Upper Saddle River, New Jersey.

over the **voltage**, **frequency**, and **duration** of the stimulation pulse. The pulse wave most commonly generated is the square wave pulse (Figure 5.2.)

The stimulator allows the application of either (+) or (−) polarity to the tissue and either a monophasic (⌒) or diphasic (⌐⌐) type of pulse wave. For most physiology experiments, a (−) polarity and a monophasic type of stimulation are employed, with a stimulation duration of 1 msec. More elaborate stimulators can deliver a pair of pulses to the tissue and have a **delay knob** that allows the control of the time interval between stimulus pairs.

Oscilloscope

The changes in electrical potential that occur during the propagation of an action potential are so small that they must be measured in units of millivolts (mV) (thousandths of a volt) or even microvolts (μV) (millionths of a volt). In addition, the speed with which these changes take place is so fast that it must be measured in units of milliseconds (thousandths of a second). To measure such small and rapid changes, highly sensitive instruments, such as the oscilloscope, must be used. The oscilloscope is an instrument that plots variations in electrical potential against time. Its component parts and the accompanying circuitry for recording the nerve compound action potential are shown in Figure 5.3. The oscilloscope is similar to a television set in that its essential component is the cathode ray tube. Within the evacuated tube, a narrow

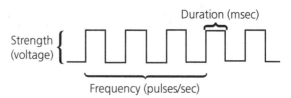

FIGURE 5.2 Square wave stimulation pulse.

beam of electrons is ejected from a rear-heated cathode and strikes the phosphor coating of the oscilloscope face, where it creates a glowing spot. The electron beam passes between two pairs of plates arranged at right angles to each other.

An electrical potential applied to one pair of plates moves the electron beam up or down. The input signal will be applied to these plates and, thus, the vertical trace indicates the magnitude of the signal being received from the biological source being recorded. An internal electrical circuit, called the **sweep circuit** or time base circuit, regularly applies a varying potential to the second pair of plates. This varying potential causes the beam to move from left to right and can be regulated to move the beam at a precise speed.

Digital storage oscilloscopes convert analog signals to digital information prior to display on a screen. Data are sampled at discrete intervals, stored, and then displayed when enough information to construct a waveform has been accumulated. Because data are stored, transient signals can be preserved and analyzed after the event has occurred. Most models can perform a number of automated or screen-based measurements.

CLINICAL APPLICATION

Nerve and Muscle Function Tests

Skeletal muscle contraction can be impaired by problems with the muscle itself or with the nerves stimulating those muscles. Diagnostic tests such as an electromyogram evaluate muscle function by examining the electrical activity in muscles that are at rest and contracting. A nerve conduction study measures the speed and strength of nerve conduction. In this test, a health professional places a shock-emitting electrode directly over the nerve to be studied and a recording electrode over the muscles supplied by that nerve. The shock-emitting electrode sends repeated, brief electrical pulses to the nerve, and the recording electrode records the time it takes for the muscle to contract in response to the electrical pulse. These tests are often given together, and the results allow a physician to determine whether a muscle is diseased, as seen in muscular dystrophy, or whether nerve function is impaired, as seen in amyotrophic lateral sclerosis (ALS, known as Lou Gehrig's disease). Ailments such as carpal tunnel syndrome, in which compression of the

Copyright © 2015, 2011, 2008 Pearson Education, Inc.

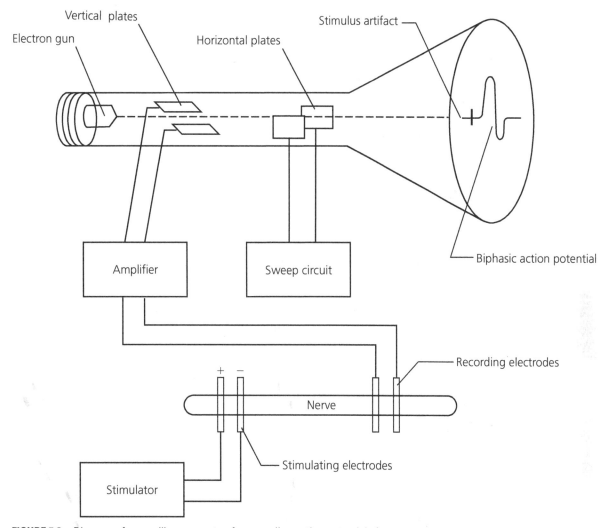

FIGURE 5.3   Diagram of an oscilloscope setup for recording action potentials from nerves.

median nerve causes numbness and weakness in the hand, can be treated by surgeries that cut the flexor retinaculum (the transverse carpal ligament surrounding the median nerve) to reduce the compression of the nerve.

▶ ACTIVITY 5.1

Sciatic Nerve Compound Action Potential

Materials

☐ 1 oscilloscope

☐ 1 stimulator

☐ 1 nerve chamber

☐ 2 glass probes

☐ 1 dissecting needle

☐ 1 10-cm piece of cotton string

☐ 1 each 250-ml squeeze bottle of frog Ringer's solution at room temperature, 10°C, and 35°C

☐ 10 ml 0.3 M KCl solution

☐ 1 set Gaskell clamps

☐ 10 ml 50% ethyl alcohol

☐ Dissection equipment (forceps and scissors)

1. Setup

Connect the stimulator, oscilloscope, and nerve chamber as shown in Figure 5.4. Double pith a large frog and dissect out (using glass probes) the sciatic nerve, which extends from the spinal cord to its insertion into the gastrocnemius muscle.

⚠ CAUTION!

Use extreme care to avoid damaging the nerve by touching it with dry fingers or metal probes or by excessive stretching.

Copyright © 2015, 2011, 2008 Pearson Education, Inc.

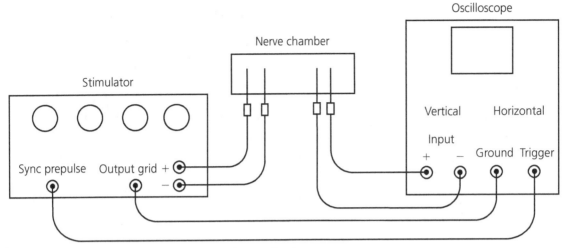

FIGURE 5.4 Connected stimulator, oscilloscope, and nerve chamber.

Your success in this activity requires that the nerve not be damaged or depolarized excessively. Place the nerve in a beaker of aerated frog Ringer's solution and allow it to equilibrate for 30 to 60 min.

EXPLAIN THIS!

In terms of normal body function, describe how Ringer's solution helps in the function of the nerve, and use this information to answer the question "What normal tissue of the body does Ringer's solution replace?"

---

2. Oscilloscope and Stimulator Controls

Familiarize yourself with the oscilloscope and stimulator controls. Your instructor will demonstrate the major controls. Keep in mind that the settings provided may need to be adjusted to obtain the best results.

3. Calibration of Oscilloscope

The vertical and horizontal sensitivities of the oscilloscope may both be calibrated either by using an internal voltage provided by the oscilloscope or by using the stimulator square wave pulse. To use the internal calibration, disconnect the lead wires from the nerve chamber to the oscilloscope and connect a lead from the calibration outlet on the scope to either the (+) or the (−) of the vertical channel. This connection will put a 1-V square wave pulse into the scope for use in calibration of the trace.

To use the stimulator pulse for calibration, connect the stimulator (+) and the output (−) terminals directly to the vertical channel input terminals of the oscilloscope (bypassing the nerve chamber). Using the stimulator controls, vary the amplitude and duration of the square wave and adjust

the vertical sensitivity and horizontal sweep speed to the desired calibration.

4. Stimulus Artifact

In the stimulus artifact activity, you will use an artificial or "shirttail" nerve to demonstrate how stimulus artifact voltage looks on the oscilloscope. This voltage sometimes resembles the compound action potential generated by the nerve and thereby confuses the observer.

Lay a thread that has been thoroughly soaked in Ringer's solution across the stimulating and recording electrodes in the nerve chamber. Place a wet towel at the bottom of the nerve chamber to keep the air in the chamber saturated. Stimulate the shirttail nerve with a pulse of 1 msec duration, increasing the voltage until the stimulus artifact appears on the oscilloscope screen. Vary the vertical sensitivity and sweep speed of the scope so that you become familiar with the shape of this artifact and can recognize it at different scope settings.

5. Threshold, Submaximal, and Maximal Stimuli

Remove the shirttail nerve and carefully lay the sciatic nerve across the electrodes in the nerve chamber. Lift the nerve only by the threads at each end. Be careful not to stretch it. Apply Ringer's solution to the nerve to ensure good conduction between nerve and electrodes.

1. Set the oscilloscope vertical sensitivity to 2 mV per division and the sweep speed to 1 msec per division.
2. Stimulate the nerve with pulses of 0.1 msec duration at a frequency of 15 to 30/sec. Gradually increase the stimulus voltage until the compound action potential appears. If the potential is not stationary on the oscilloscope,

Copyright © 2015, 2011, 2008 Pearson Education, Inc.

alter the sweep speed until a standing wave is obtained. If you are using a digital oscilloscope, the prepulse signal from the stimulator can be connected to the channel 2 input on the oscilloscope. You can then set the trigger source to channel 2 and obtain a very stable waveform. Distinguish between the stimulus artifact and the action potential. Determine the threshold voltage for the most sensitive nerve fibers and **record it in the Laboratory Report.**

3. Now continue increasing the stimulating voltage gradually, observing the changes in the action potential as you do. You will eventually reach a point (the maximal stimulus voltage) at which the action potential does not increase further (at maximal stimulus).

Record your observations in the Laboratory Report, using the following questions as a guide:

Does the sciatic nerve follow the all-or-none law of neurons?

What is the maximal height of the action potential in millivolts?

What is its duration in milliseconds?

6. Strength–Duration Curve

The ease with which a membrane can be stimulated depends on two variables: the strength of the stimulus and the duration for which the stimulus is applied. These variables are inversely related: As the strength of the applied current increases, the time required to stimulate the membrane decreases

1. Determine the threshold voltage of the nerve at a variety of pulse durations (e.g., 0.02, 0.03, 0.04 msec). Use as the threshold response either the first appearance of the nerve action potential or half the maximal height of the action potential (when approximately half of the neurons have reached threshold). **Tabulate the data and plot a strength–duration curve in the Laboratory Report.**

2. From this plot, determine the **rheobase voltage** and **chronaxie time** for the nerve membrane. **Rheobase voltage** is the lowest voltage capable of reaching threshold and producing an action potential, no matter how long the duration of the stimulus. **Chronaxie** is the minimum **time** required for an electric current double the strength of the rheobase to stimulate a neuron. How are strength and duration of stimulus related in obtaining a threshold response? What is the significance of the chronaxie time? **Enter your observations in the Laboratory Report.**

7. Conduction Velocity

1. Stimulate the nerve with a maximal stimulus voltage to obtain a full compound action potential. Adjust the delay knob on the stimulator so the stimulus artifact can be seen clearly. If your stimulator does not have a delay knob, use the origin of the oscilloscope trace as the beginning of the stimulus artifact.

2. Determine the speed of conduction of the action potential over the nerve as follows:

   a. Measure the distance (D) in millimeters from the (−) stimulating electrode and the (+) recording electrode in the nerve chamber.

   b. Determine the time (T) in milliseconds between the stimulus artifact and the beginning of the compound action potential. To obtain this, measure the distance between the stimulus artifact and the beginning of the compound action potential in centimeters and multiply it by the sweep speed of the oscilloscope in milliseconds per centimeter.

   c. Calculate the conduction velocity, using the following formula:

   Impulse velocity (mm/msec or m/sec)

   $$= \frac{\text{Distance}(D)(\text{mm})}{\text{Time}(T)(\text{msec})}$$

How does the conduction velocity you obtained compare with published values for the sciatic nerve (around 25 to 35 m/sec)? **Enter your observations in the Laboratory Report.**

8. Refractory Period

1. Stimulate the nerve using maximal voltage of 1-msec duration at a frequency of 10 pulses per second.

2. Change the stimulator setting to twin pulses instead of regular. In this mode, the nerve is stimulated by twin pulses, with the second pulse following the first pulse by the time (in msec) indicated on the delay knob of the stimulator. When the delay time is of sufficient length, you will see two full-size compound action potentials on the scope.

3. Now begin slowly reducing the stimulus delay time, and you will see the second potential gradually move toward the first potential. As it nears the first potential, its height decreases (why?) until the entire second potential disappears. Measure the distance on the scope from the point of disappearance of the second potential to the beginning of the first potential and, from this, determine the refractory period (in msec) of the nerve.

Copyright © 2015, 2011, 2008 Pearson Education, Inc.

What mechanism is responsible for the refractory period? What is the significance of this period? **Enter your observations in your Laboratory Report.**

9. Summation of Subliminal Stimuli

1. Using a short pulse duration (e.g., 0.01 to 0.1 msec), determine the threshold voltage for your nerve. Stimulate the nerve with twin pulses, using a voltage that is slightly below threshold (subliminal). Decrease the delay time between the pulses until threshold is reached and an action potential is produced. Summation of subliminal stimuli is sometimes difficult to obtain, and you may have to try several combinations of duration and voltage until you are successful. **Enter your observations in your Laboratory Report.**

10. Direction of Propagation

1. Reverse the stimulating and recording electrodes or turn the nerve around. Stimulate the nerve. Is the impulse conducted in the opposite direction? How can you reconcile this result with the concept of unidirectional propagation between neurons? **Enter your observations in your Laboratory Report.**

11. Monophasic Action Potential and Nerve Fiber Types

1. Place a drop of isotonic potassium chloride (KCl) solution on, or crush (using forceps), the portion of the nerve lying over the distal recording electrode. Gradually increase the stimulus voltage from threshold to maximal and note the order of appearance of the various peaks, or elevations, in the action potential.

What is the threshold for each of these peaks? What do the peaks represent? Why are these elevations seen only when a monophasic potential is produced? What is meant by a compound action potential? **Enter your observations in your Laboratory Report.**

12. Nerve Conduction Blockade

The propagation of nerve impulses can be blocked by many agents, such as the local anesthetics, procaine and lidocaine (Xylocaine), and the general anesthetics, ether and chloroform. Even alcohol can depress nerve conduction. Of the different nerve types, the fibers that have relatively small diameters are the most sensitive to anesthetic action, and larger fibers are the most resistant. Thus, in a mixed nerve, sensory neurons are the first to be anesthetized, and motor neurons

are the last. Nerve conduction can also be blocked by local application of pressure greater than 130 mm Hg. With pressure blockade, however, the sensitivity picture is reversed—motor fibers are blocked before sensory fibers are affected.

a. Drug Blockade
   Dampen a small piece of cotton with 50% ethanol, squeeze out the excess fluid, and place the cotton on the nerve between the stimulating and the recording electrodes. Avoid touching the nerve with your fingers. Cover the chamber. Stimulate the nerve continuously and observe the action potential for several minutes. What happens? Is there a sequence of changes? Remove the cotton before complete nerve block occurs and apply fresh frog Ringer's solution to the nerve. Is recovery complete?

b. Pressure Blockade
   Cover the jaws of a Gaskell clamp with rubber tubing and moisten the tubing with Ringer's solution. Place the clamp over the nerve between the stimulating and recording electrodes. Stimulate the nerve continuously and gradually apply pressure by closing the jaws of the clamp. Apply pressure until conductivity is completely blocked. Release the pressure and test to determine whether nerve conduction returns. **Enter your observations in your Laboratory Report.**

5.1 POWERLAB® VERSION

Materials needed for this activity will have been set up for you.

Materials

☐ 1 computer

☐ 1 PowerLab® data acquisition unit

☐ 1 nerve chamber

☐ 2 glass probes

☐ 1 dissecting needle

☐ 1 10-cm piece of cotton string

☐ 1 each 250-ml squeeze bottle of frog Ringer's solution at room temperature, 10°C, and 35°C

☐ 10 ml 0.3 M KCl solution

☐ 1 set Gaskell clamps

☐ 10 ml 50% ethyl alcohol

☐ Dissection equipment (forceps and scissors)

⚠ CAUTION!

Use extreme care to avoid damaging the nerve by touching it with dry fingers or metal probes or by excessive stretching.

Copyright © 2015, 2011, 2008 Pearson Education, Inc.

Your success in this activity requires that the nerve not be damaged or depolarized excessively. Place the nerve in a beaker of aerated frog Ringer's solution and allow it to equilibrate for 30 to 60 min.

EXPLAIN THIS!

In terms of normal body function, describe how Ringer's solution helps in the function of the nerve, and use this information to answer the question "What normal tissue of the body does Ringer's solution replace?"

---

1. Setup

a. Launch **Scope**.

b. The recording electrodes from the nerve box have been connected to channel 1. Accordingly, make sure that data collected from channel 1 are displayed on Input A (the top right of the window).

c. Choose **Off** for Input B.

d. You will be stimulating the nerve using the PowerLab® built-in stimulator. Under **Setup** in the Scope window, choose **Stimulator**. A pop-up window (Figure 5.5) will allow you to adjust the mode (Off, Pulse, Multiple), delay (when the stimulus will be delivered after you start recording data), duration (in ms), and amplitude (in mV) of the stimulus. Make sure that the Isolated Stimulator box is cleared.

e. Set mode to Pulse; delay to 2,500 msec; duration to 1 msec; and amplitude to 0.1 mV.

f. In the Stimulator window, a vertical line will appear in the graph at time 2,500 msec, representing the amplitude of the stimulus. The height of this line will change, depending upon the amplitude of the stimulus. Likewise, the position of the line will change, depending on where you set the delay.

g. Click **OK**.

h. Under Display, click **Overlay stimulator**. In this window, set mode to vertical lines and color to red or any other color. This setting will overlay the stimuli over the response of the recording electrodes, allowing you to easily visualize the response of the nerve to each stimulus.

i. Click **OK**.

j. To the right of the screen, you will see drop-down menus for Input A and Input B. Set Input A to channel 1 and range to 10 mV. Input B should be off.

k. Under Time Base, set samples to 512 and time to 5 sec.

l. Click **Start** to test the setup.

2. Stimulus Artifact

In the stimulus artifact activity, you will use an artificial, or shirttail, nerve to demonstrate how stimulus artifact voltage looks on the oscilloscope. This voltage sometimes resembles the compound action potential generated by the nerve and thereby confuses the observer.

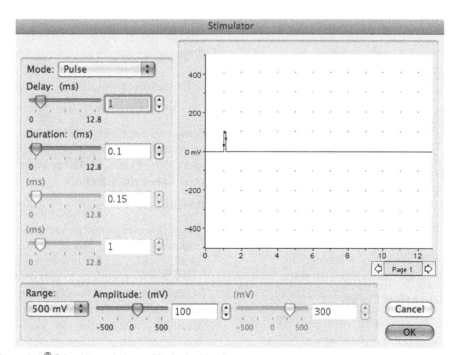

FIGURE 5.5 PowerLab® Stimulator window with single stimulus.

Copyright © 2015, 2011, 2008 Pearson Education, Inc.

a. Lay a thread that has been thoroughly soaked in Ringer's solution across the stimulating and recording electrodes in the nerve chamber. Place a wet towel at the bottom of the nerve chamber to keep the air in the chamber saturated.

b. Stimulate the shirttail nerve with a pulse of 1 msec duration, increasing the voltage until the stimulus artifact appears on the oscilloscope screen. Vary the range and time (as described in the preceding steps j and k) so that you become familiar with the shape of this artifact and can recognize it at different settings.

3. Threshold, Submaximal, and Maximal Stimuli

a. Remove the shirttail nerve and carefully lay the sciatic nerve across the electrodes in the nerve chamber. Lift the nerve only by the threads at each end. Be careful not to stretch it. Apply Ringer's solution to the nerve to ensure good conduction between nerve and electrodes.

b. Set the range to 5 mV and time to 10 msec. Use the Stimulator menu to deliver pulses of 0.1 msec duration, delay to 1 ms, and amplitude of 0.50 mV. Click **OK** to close the Stimulator window and then click **Start**. Scope will stimulate the nerve and record the response in the active window. Repeat the process while gradually increasing the voltage of the stimulus by 50 mV until the compound action potential appears. Scope will save a new page each time you click Start.

c. Distinguish between the stimulus artifact and the action potential. Determine the threshold voltage for the most sensitive nerve fibers and **record it in the Laboratory Report**. Now continue increasing the stimulating voltage gradually, observing the changes in the action potential as you do. You will eventually reach a point (maximal stimulus voltage) at which the action potential does not increase further (at maximal stimulus).

Does the sciatic nerve follow the all-or-none law of neurons? What is the maximal height of the action potential in millivolts? **Enter your observations in your Laboratory Report.**

4. Strength–Duration Curve

a. Using the Stimulator menu, determine the threshold voltage of the nerve at a variety of pulse durations (e.g., 0.02, 0.03, 0.04 msec).

Use as the threshold response either the first appearance of the nerve action potential or half the maximal height of the action potential (when approximately half of the neurons have reached threshold). **Tabulate the data and plot a strength–duration curve in the Laboratory Report.**

b. From this plot, determine the **rheobase voltage** and **chronaxie time** for the nerve membrane. **Rheobase voltage** is the lowest voltage capable of reaching threshold and producing an action potential, no matter how long the duration of the stimulus. **Chronaxie** is the minimum **time** required for an electric current double the strength of the rheobase to stimulate a neuron. How are strength and duration of stimulus related in obtaining a threshold response? What is the significance of the chronaxie time? **Enter your observations in your Laboratory Report.**

5. Conduction Velocity

Stimulate the nerve with a maximal stimulus voltage to obtain a full compound action potential. Determine the speed of conduction of the action potential over the nerve as follows:

a. Measure the distance (D) in millimeters from the (−) stimulating electrode and the (+) recording electrode in the nerve chamber.

b. Use the cursor to determine the time (T) in milliseconds between the stimulus artifact and the beginning of the compound action potential.

c. Calculate the conduction velocity, using the following formula:

Impulse velocity (mm/msec or m/sec)

$$= \frac{Distance(D)(mm)}{Time(T)(msec)}$$

How does the conduction velocity you obtained compare with published values for the sciatic nerve (around 25 to 35 m/sec)? **Enter your observations in your Laboratory Report.**

6. Refractory Period

a. Under the Stimulator menu, make sure that the Isolated Stimulator box is cleared. Set mode to Double, delay to 100 msec, and duration to 1 msec. Set amplitude A and B to the maximal stimulus voltage. A sample Stimulator window is shown in Figure 5.6.

b. Gradually increase the interval between pulses to determine the refractory period of the nerve. In this mode, the nerve is stimulated by

Copyright © 2015, 2011, 2008 Pearson Education, Inc.

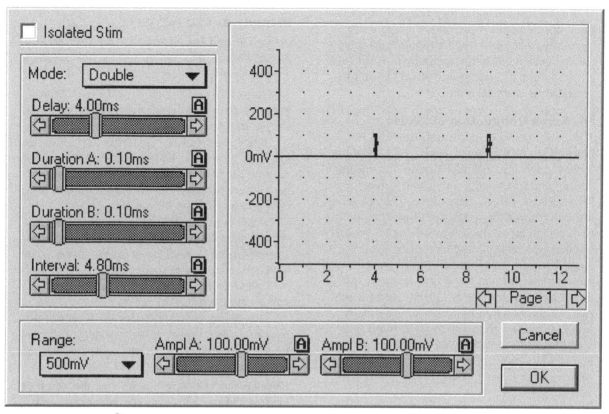

FIGURE 5.6   PowerLab® Stimulator window with dual pulses.

twin pulses, the second pulse following the first pulse by the time (in msec) indicated by the interval of the stimulator menu. When the interval time is of sufficient length, you will be able to see two full-size compound action potentials on the scope.

c. Now begin slowly reducing the stimulus interval time, and you will see the second potential gradually move toward the first potential. As it nears the first potential, its height decreases (why?) until the entire second potential disappears. Use the cursor to determine the point of disappearance of the second potential to the beginning of the first potential, and from this, determine the refractory period (in msec) of the nerve.

What mechanism is responsible for the refractory period? What is the significance of this period? **Enter your observations in your Laboratory Report.**

7. Summation of Subliminal Stimuli

a. Using a short pulse duration (e.g., 0.01 to 0.1 msec), determine the threshold voltage for your nerve. Stimulate the nerve with twin pulses, using a voltage that is slightly below threshold (subliminal).

b. Decrease the delay time between the pulses until threshold is reached and an action potential is produced. Summation of subliminal stimuli is sometimes difficult to obtain, and you might have to try several combinations of duration and voltage until you are successful. **Enter your observations in your Laboratory Report.**

8. Direction of Propagation

Reverse the stimulating and recording electrodes or turn the nerve around. Stimulate the nerve using maximal stimulus voltage. Is the impulse conducted in the opposite direction? How can you reconcile this result with the concept of unidirectional propagation between neurons? **Enter your observations in your Laboratory Report.**

9. Monophasic Action Potential and Nerve Fiber Types

a. Place a drop of isotonic potassium chloride (KCl) solution on, or crush (using forceps), the portion of the nerve lying over the distal recording electrode.

b. Gradually increase the stimulus voltage from threshold to maximal and note the order of appearance of the various peaks, or elevations, in the action potential.

Copyright © 2015, 2011, 2008 Pearson Education, Inc.

What is the threshold for each of these peaks? What do the peaks represent? Why are these elevations seen only when a monophasic potential is produced? What is meant by a compound action potential? **Enter your observations in your Laboratory Report.**

10. Nerve Conduction Blockade

The propagation of nerve impulses can be blocked by many agents, such as the local anesthetics, procaine and lidocaine (Xylocaine), and the general anesthetics, ether and chloroform. Even alcohol can depress nerve conduction. Of the different nerve types, the fibers that have relatively small diameters are the most sensitive to anesthetic action, and larger fibers are the most resistant. Thus, in a mixed nerve, sensory neurons are the first to be anesthetized, and motor neurons are the last. Nerve conduction can also be blocked by local application of pressure greater than 130 mm Hg. With pressure blockade, however, the sensitivity picture is reversed—motor fibers are blocked before sensory fibers are affected.

a. Drug Blockade
   Dampen a small piece of cotton with 50% ethanol, squeeze out the excess fluid, and place the cotton on the nerve between the stimulating and the recording electrodes. Avoid touching the nerve with your fingers. Cover the chamber. Stimulate the nerve continuously and observe the action potential for several minutes. What happens? Is there a sequence of changes? Remove the cotton before complete nerve block occurs and apply fresh frog Ringer's solution to the nerve. Is recovery complete?

b. Pressure Blockade
   Cover the jaws of a Gaskell clamp with rubber tubing and moisten the tubing with Ringer's solution. Place the clamp over the nerve between the stimulating and recording electrodes. Stimulate the nerve continuously and gradually apply pressure by closing the jaws of the clamp. Apply pressure until conductivity is completely blocked. Release the pressure and test to determine whether nerve conduction returns. **Enter your observations in your Laboratory Report.** ◀

INQUIRY-BASED ACTIVITY

Appendix C describes the format of a typical Laboratory Report. It is mandatory that you read the Appendix before you start planning your experiment.

The methodology described above can be used to test the effects of temperature or alcohol concentration on the experiments described in activities 5.1.3 through 5.1.8.

If testing temperature effects, make sure that you use frog Ringer's solution to either cool or warm the nerve in the nerve box. Make sure that you start at the lowest temperatures and proressively use warmer solutions. Allow the nerve to recover between observations.

If testing the effects of various concentrations of alcohol, use the methodology described in 5.1.12 to make your observations. Start with the lowest concentration of alcohol and progressively increase the concentration of the alcohol used. Allow the nerve to recover between obervations by spraying it with frog Ringer's solution.

Analyze your data to determine the effect that temperature or alcohol have on the threshold, size, and shape of the action potential, the conduction velocity and the refractory period.

Complete this checklist to make sure that you have covered all bases before you start your experiment:

What is the question you seek to answer?

Frame it in the form of a hypothesis.

What is the independent variable or treatment?

How will the independent variable or treatment be measured?

What is the control treatment?

How will you replicate your experiment?

How will you ensure that your subjects are similar enough to not introduce some other independent variability? Are there any standardized variables?

What is/are the dependent variable(s)?

How will the dependent variable(s) be measured?

What are your predictions?

Construct a table to record your observations easily.

How will you present the data collected graphically?

How will you analyze the data collected?

How will you know if the differences between the treated and the untreated samples are statistically significant?

Your laboratory instructor will describe how your Laboratory Report will be written.

Copyright © 2015, 2011, 2008 Pearson Education, Inc.

Reflex Functions

[handwritten: reaction time]

[handwritten notes in margin:]
hearing reaction
750 ms   677 ms
671   627 ms
691   664 ms
675   715 ms
686   656 ms

[handwritten:]
1) 456 ms   267
2) 365 ms   385
3) 321 ms   271
4) 305 ms   338
5) 307 ms   344

CHAPTER 6 INCLUDES:

BIOPAC Systems, Inc.   1 BIOPAC® Activity

OBJECTIVES

After completing this exercise, you should be able to

1. Describe the nervous pathways involved in a reflex action.
2. Demonstrate the physiology of eye reflex actions and spinal reflex actions.
4. Explain the physiology of the labyrinthine reflexes and their effect on balance.
5. Explain the physiology of reflexes affecting reaction responses.

Human Reflexes

The essentials of a reflex mechanism are a receptor organ, an effector organ, and some type of communications network connecting the two. Reflex action is initiated by an input stimulus and results in an output response. Reflex activity ranges from the simple axon reflex to the complex reflexes in which the cerebrum participates.

Many reflexes might be regarded as being programmed; that is, the appropriate response to the stimulus has been built into the nervous system. The spinal reflexes that require transmission from the periphery to the spinal cord and then back to the appropriate effector organ are examples of this kind of programming. For instance, when one experiences a painful stimulus such as burning a finger on a hot object, the spinal reflex immediately causes withdrawal of the finger from the offending object. No action is required by the central nervous system. The reflex functions equally well in an animal whose spinal cord has been divided above the location of the cell bodies of the participating nerves.

Other reflexes, such as eye reflexes and labyrinthine reflexes, require action of centers in the brain. In these instances, the appropriate response is determined after several different inputs have been evaluated; hence, integrative function of the central nervous system is required.

In the following activities, you will investigate several types of human reflexes to demonstrate their integrative function at several levels of integration in the body.

▶ ACTIVITY 6.1
Eye Reflexes

Materials (per group)

☐ 1 flashlight

1. Pupillary Reflex *[handwritten: pupils flashing]*
Observe the size of your partner's pupils in a given intensity of light. Flash a light into one eye and observe the pupillary responses. **Enter your observations in the Laboratory Report.**

Observe the diameter of the pupils in a given light. Without changing either the light intensity or the focus, place your hand over one eye. **Enter your observations in the Laboratory Report.**

2. Accommodation Reflex *[handwritten: got smaller]*
Observe the size of your partner's pupils when the eyes are focused on a distant object (more than 20 ft away). Watch carefully while the focus is shifted to a near object. Do not change the light intensity. **Enter your observations in the Laboratory Report.**

3. Corneal (Blink) Reflex
Move your hand suddenly toward your partner's eyes. **Enter your observations in the Laboratory Report.** ◀

▶ ACTIVITY 6.2
Spinal Reflexes

Materials (per group)

☐ 1 reflex hammer

Copyright © 2015, 2011, 2008 Pearson Education, Inc.

1. Patellar Reflex

Have your partner sit in a chair or on a laboratory stool with legs crossed. Be sure to maintain a relaxed lower body. Gently tap the patellar tendon of the crossed leg with a reflex hammer and note the response. Compare responses of the right and left knees. **Diagram the reflex arc in the Laboratory Report.**

Repeat the process but instruct your partner to perform the **Jendrassik maneuver**, that is, to clasp his or her hands in front and, with fingers locked, try vigorously to pull the hands apart at the same time that the tendon is tapped. How do you explain the responses obtained? **Enter your explanation in the Laboratory Report.**

2. Achilles Reflex

Have your partner kneel on a chair or stool with the feet hanging free and relaxed over the edge. Bend his or her foot downward to increase tension on the gastrocnemius muscle. Tap the Achilles tendon lightly with a reflex hammer or the side of your hand. The contraction of the gastrocnemius causes plantar extension of the foot. Have your partner grasp the back of the chair and repeat. **Enter your observations in the Laboratory Report.**

3. Biceps and Triceps Reflexes

Place your forearm on the laboratory table so that the elbow is bent to approximately 90 degrees; push your forearm vigorously against the table top. Palpate the biceps muscle. Try to contract the biceps to flex your arm while continuing to press down against the tabletop. **Enter your observations in the Laboratory Report.**

Flex your arm to 90 degrees and attempt to lift the table. Palpate the triceps muscle. Try to contract the triceps. **Record your results in the Laboratory Report.** ◄

►ACTIVITY 6.3
Ciliospinal Reflex

Pinch the skin on one side of the nape of your neck and note the dilation of the pupil of the eye on the ipsilateral side. This is a reflex response mediated over the sympathetic nervous system in response to a painful stimulus. ◄

►ACTIVITY 6.4
Plantar Reflex and Babinski's Reflex

Keeping your lower body relaxed, have your partner scratch or sharply stroke the sole of your foot near the inner side, using a blunt probe. The normal adult reflex response is a plantar (downward) flexion or curling of all toes. If the toes fan out with the big toe flexed dorsally (upward), the response is referred to as the positive Babinski reflex; this reflex is often associated with damage to the pyramidal tract fibers. Babinski's reflex is the normal response of a child in its first year because the nerves are still undergoing myelination at this time. ◄

►ACTIVITY 6.5
Labyrinthine Reflexes

Materials (per group)

☐ 1 swivel armchair with casters removed

The purpose of this activity is to show the role of the labyrinths in orientation of the body during movement and how labyrinthine reflexes correlate to muscular movements and eye movements reflexly to maintain equilibrium.[1]

Angular acceleration during rotation is detected by the cristae ampullaris receptors in the semicircular canals. These receptors contain hair cells with stereocilia and kinocilia embedded in a gelatinous mass (cupula) that extends into the endolymph in the semicircular canals. When the stereocilia in each hair cell are bent by endolymph movement toward the kinocilium, the hair cell is depolarized, increasing the firing of the afferent nerve fibers. Conversely, when the stereocilia in the hair cell are bent away from the kinocilia, the hair cell is hyperpolarized, reducing the firing of the afferent nerve fibers.

During rotation of the horizontal canal to the right, the inertia of the endolymph causes it to lag behind the canal movement. The relative movement of the endolymph to the left displaces the cupula, thus bending the hair cell filaments and altering the nerve activity to the brain (Figure 6.1a). This provides the brain with sensory information that the head is being accelerated. After 15 to 20 sec of rotation, when a constant velocity is reached, the

---

[1]People who are subject to motion sickness are good subjects for the labyrinthine reflex tests because they show a strong response. Because of the discomfort they might experience, however, these persons should be subjected to only one test in which they are rotated.

Copyright © 2015, 2011, 2008 Pearson Education, Inc.

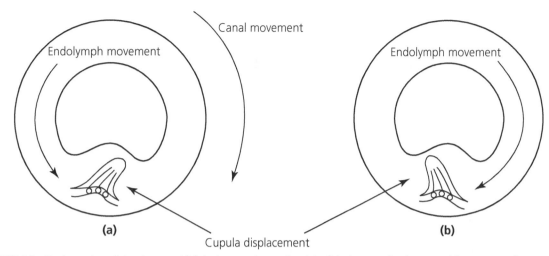

Canal movement

Endolymph movement

Endolymph movement

Cupula displacement

(a)   **(b)**

FIGURE 6.1   Horizontal semicircular canal (a) during rotation to the right; (b) after rotation is stopped (postrotatory).

endolymph movement catches up with the canal movement. The cupula then assumes its normal upright position, and the sensation of acceleration is lost.

After rotation is stopped, the canal is stationary, but the endolymph continues moving to the right, due to its inertia. This postrotatory endolymph movement displaces the cupula in the opposite direction (Figure 6.1b); this modifies the flow of nerve impulses to the brain, which produces a sensation of spinning to the left. Because the semicircular canals on each side of the head respond differently to different types of movement, we are instantly informed of our position in space as we move.

Through reflex neural connections, the cristae ampullaris receptors are linked to centers in the brain that control eye movements and muscle tone. These relationships are readily demonstrated in the following activities.

1. Nystagmus   → left

One of the functions of the semicircular canal mechanism is to aid in visual fixation on moving targets. If the canals are stimulated under experimental conditions, reflex responses result in movement of the eyes, called **nystagmus**. Nystagmus has two components, a fast and a slow phase. The *direction* of nystagmus is designated as that of the fast phase. If a person is angularly accelerated in one direction, the eyes will move very slowly in one direction, as though to maintain fixation on a moving target, and then very rapidly swing back in the other direction. This action is *rotary* nystagmus and is caused by the acceleration of fluid in the semicircular canals, which stimulates the crista in the ampulla and produces the sensation of turning. It is very difficult to observe rotary nystagmus, but the same

phenomenon can be obtained by suddenly stopping a spinning person and observing *postrotary* nystagmus. In observing postrotary nystagmus, however, you must remember that all movements have been reversed and the direction of nystagmus is reversed. It is important to read the directions below completely before starting the procedure. Please notice that *all* rotations performed will be to the right (clockwise).

a. Seat the subject in an office swivel armchair from which the casters have been removed. Several members of the group should stand nearby to catch the subject in case he or she topples from the chair.

b. With the subject's feet off the floor and head bent forward at an angle of 30 degrees, rotate the subject to the right at the rate of one rotation every 2 sec for 10 turns. Stop the rotation suddenly and observe the motion of the eyes at the moment the movement of the chair stops. Note the direction of nystagmus.

c. Repeat the rotation with the head bent to the right shoulder at an angle of about 90 degrees. What is the direction of nystagmus? **Enter your observations in the Laboratory Report.**

2. Past Pointing   → right

With his or her eyes open, have the subject extend an arm to touch the finger of the operator; then with the eyes closed, have the subject repeat the process. Most people can perform this demonstration without difficulty.

Now rotate the subject to the right as before. Stop the rotation abruptly and let the subject reach out, as previously, to touch the operator's finger. Let the subject immediately close his or her eyes and try to touch the finger.

Copyright © 2015, 2011, 2008 Pearson Education, Inc.

In which direction did the subject err in relation to the direction of rotation? How do you explain the direction of pointing? **Enter your explanation in the Laboratory Report.**

3. Equilibrium → left †

With the subject's head bent forward at an angle of 30 degrees, rotate him or her to the right as before. Immediately after cessation of rotation, have the subject attempt to rise and walk. Observe in which direction the subject appears to fall. Ask the subject to describe the sensation experienced as he or she attempts to walk after discontinuation of rotation. **Enter your observations in the Laboratory Report.**

Repeat the procedure, but with the head bent to the right shoulder at an angle of about 90 degrees. Record your observations and explain. **Enter your explanation in the Laboratory Report.**

4. Proprioception and Spatial Orientation

Have the subject hold his or her arms stretched forward with index fingers pointed toward each other and then try to bring the fingertips together. Test this reaction first with the subject's eyes open and then with the eyes closed. Record the results.

Have the subject stand with feet together and arms outstretched. Observe body sway and the corrective motions required to maintain balance and then repeat the test, first with the subject's eyes open and then with the eyes closed. Then have the subject stand first on one foot and then on the other. Again perform the test with the eyes closed and open. These tests evaluate the static equilibrium of the person and point out the contribution made by the eyes to static equilibrium.

Have the subject look at the ceiling and stand on one foot. Again have the subject stand on one foot, with his or her head in the same position as before, but this time with the eyes closed.

Reaction Times

The time required for a person to react to a stimulus depends on several factors: the responsiveness of the receptor, nerve conduction velocity, synaptic delay, number of neurons and synapses involved, nerve distance to be traveled, the efficiency of neuromuscular transmission, and the speed of muscle contraction. The type of response is also important. In an automatic reflex, relatively few synapses are traversed and the response time is short. In a response that requires thought, decision making, and choice, more neural pathways must be traversed and the response time is longer. In the following

activity, you will measure and compare the times required to complete various reaction responses.

STOP AND THINK

It was once thought that alcohol's depressive effect was primarily due to the action of alcohol as a membrane disruptor, resulting in a broad effect on the brain due to the ability of the small alcohol molecules to cross the blood-brain barrier. It is now known that this effect is due to alcohol targeting gamma-aminobutyric acid (GABA) receptors and acting as an agonist of this major inhibitory neurotransmitter in the brain. When coupled with alcohol inhibiting the major excitatory NT, glutamate, and releasing other inhibitors such as dopamine and serotonin, it is easy to explain the effect of alcohol on our reaction times.

The reaction time apparatus consists of the operator's initiate console, the subject's response panel, and a stop clock. The operator's console and the clock should be placed as far as possible from the subject's panel so the subject cannot see the operator's console. When the operator depresses the initiate button, the stop clock and the stimulus (colored lamp or buzzer) are activated simultaneously. When the subject depresses the proper response button, the clock is stopped and the reaction time can be read on the clock in hundredths or thousandths of a second. The operator can select the buzzer sound or any one of four colors (red, white, blue, or green) by turning the knob on the operator's console.

In each of the following tests, obtain a minimum of three measurements, calculate the average reaction time for each test, and record your results in the Laboratory Report. If the subject responds prematurely, disregard that particular time. ◄

►ACTIVITY 6.6

Reaction Times

Materials (per group)

☐ 1 reaction time apparatus

☐ 1 stopwatch

1. Sight

a. Test the subject's reaction time to a single color stimulus, using one response button.

b. Determine the reaction time to a choice of two colors, using two response buttons (with a specified response button for each color).

c. Determine the reaction time to a choice of three colors, using three response buttons.

Copyright © 2015, 2011, 2008 Pearson Education, Inc.

d. Determine the reaction time to a choice of four colors, using four response buttons. **Enter your observations in the table in the Laboratory Report.**

2. Sound

Test the subject's reaction time to the buzzer sound produced when the initiate button is depressed.

3. Word Association

In this test, when you depress the initiate button, simultaneously give a stimulus word—for example, "bread." The subject will reply with a word he or she associates with the stimulus word (e.g., "butter") and will simultaneously depress a previously designated response button. Record the stimulus and response words used in each test and the reaction time required. You will see quite a variation in time depending on the stimulus words used—and on the subject. **Enter your observations in the table in the Laboratory Report.**

6.6 BIOPAC® VERSION

Materials

☐ BIOPAC® SS10L Hand switch

☐ BIOPAC® OUT1 Headphones

☐ Computer with BIOPAC® Student lab software

Materials needed for this activity will have been set up for you.

1. Double click the "Startup" file in the "Biopac Student Lab 3.x.x" folder or the desktop icon created by your instructor to start the program. A window with lesson options will appear.
2. Double click "L11 Reaction Time 1" and the settings for this experiment will be loaded. A panel requiring identifying information will appear.
3. Enter your first name and last name or some other unique identifying information in the panel and click OK.
4. Record this information because you will need it to analyze your data. Click OK once again in the box that informs you how to retrieve your data.

Calibration

1. Subject: Put on the headphones and hold the response switch in your dominant hand.
2. Subject: Close your eyes and place your thumb on the button of the response switch.
3. Director: Click the "**Calibrate**" button. The subject should press and immediately release the response switch when he or she hears a

click in the headphones. The calibration will automatically stop after 8 sec.
4. If the setup is correct, you should see a screen with a line showing an upward deflection ("pulse") at approximately 4 sec into the recording. This tells you that your equipment has been set up correctly. If no response is recorded, check your connections and recalibrate.

Experiment

Four trials will be conducted, each requiring the subject to press the button in response to a click in the headphones.

Trials 1 and 2 present stimuli at pseudo-random (1- to 10-second) intervals.

Trials 3 and 4 present stimuli at fixed intervals (every 4 seconds).

Experimental Method

Trials 1 and 2—Pseudo random stimuli

1. The subject should be comfortably seated, with eyes closed and the response switch in his or her hand. When the subject is ready, he or she should inform the director, who will then click the "**Record**" button.
2. The subject should respond to each click by pressing and releasing the push button on the response switch. An event maker will automatically be inserted at this time.
3. The recording will stop automatically after 10 clicks.
4. You can "**Redo**" the experiment by clicking that button if necessary.
5. Click "Resume" to conduct trial 2 and repeat steps 1–3.
6. Follow the steps above to conduct Trials 3 and 4—Fixed interval stimuli.
7. Click "**Done**."
8. Repeat steps 1–3 above for two trials for fixed interval stimuli.
9. Click "**Done**" when completed.
10. Answer "**yes**" in the panel that appears on screen.

If choosing the "Record from another Subject" option, remember to choose and record a unique file name. If all subjects have been tested, choose either "analyze current data file" or "analyze another data file." **Enter all data in the table in the Laboratory Report.**

You can repeat the experiment using your non-dominant hand. Do you think reaction time will be shorter for the dominant versus the non-dominant hand? ◄

Copyright © 2015, 2011, 2008 Pearson Education, Inc.

Reflex Functions

Name _____

Date _____ Section _____

Score/Grade _____

Human Reflexes

1. a. Pupillary reflexes

   Do both eyes change simultaneously? _____

   _____

   What is the receptor in this reflex? The effector? _____

   _____

   What happens to its diameter? _____

   _____

   Explain the change in the uncovered pupil. This is called the consensual reflex. _____

   _____

   b. Accommodation reflex

   How does the pupil size change? _____

   _____

   What is the advantage of this change? _____

   _____

   c. Corneal (blink) reflex

   Move your hand suddenly toward your partner's eyes. Describe your partner's reaction._____

   _____

2. Spinal reflexes

   a. Patellar reflex

   On a blank sheet of paper, diagram the reflex arc operating in the patellar reflex.
   What happens to the patellar reflex when the Jendrassik maneuver is performed at the same time
   the patella is struck? Explain the mechanism of the response. _____

   _____

   b. Achilles reflex

   What happens when the Achilles tendon is tapped with the reflex hammer? _____
   _____ A reflex occured _____

   What happens to the Achilles reflex when the subject grasps the back of the chair and is then
   tested? _____

   _____

   c. Biceps and triceps reflexes

   What part does reciprocal innervation play in the biceps and triceps reflexes? _____

   _____

   On a blank sheet of paper, diagram the reflex arc in the flexion reflex and the crossed extensor reflex.

3. Explain why the pupils dilate when skin is pinched. _____

   _____

Copyright © 2015, 2011, 2008 Pearson Education, Inc.

4. Why would a physician test the Babinski reflex in a young child? _____
_____

5. Labyrinthine reflexes

    a. Nystagmus

        What is the direction of nystagmus when the subject is stopped after rotation with the head bent forward 30 degrees? _____

        With the head bent to the right shoulder at a 90-degree angle? _____

        What practical function does the nystagmus reflex play in a person's physiological function?

        _____

    b. Past pointing

        After rotation to the right and stopping, did the subject point to the left or right of the operator's finger? _____

        How do you explain this direction of pointing based on teleology (that is, some practical purpose for the observed reflex)? _____

    c. Equilibrium

        In which direction did the subject move or fall when he or she stood up after being rotated to the right and then stopped with the head bent forward 30 degrees? _____

        With the head bent to the right 90 degrees? _____

        A person lays his or her head on the left shoulder and then is rotated to the left for 20 sec and stopped abruptly. What will be the postrotatory direction of nystagmus? _____

        _____

        _____

        If the subject is instructed to stand up immediately after rotation is stopped, what will probably happen to him or her? _____

        _____

        _____

        Explain the preceding response. Describe the receptor response, the sensation perceived by the subject, and the final effect on muscle tone. _____

        _____

        _____

        On a blank sheet of paper, diagram and label the labyrinthine receptors responsible for detecting static postural changes and angular acceleration.

    d. Proprioception and spatial orientation.

        Can the subject bring his fingers together with eyes open? _____

        With eyes closed? _____

        Can the subject maintain balance with feet together, arms outstretched, and eyes open? _____

        With eyes closed? _____

        Can the subject stand on either foot alone, with eyes open? _____

        With eyes closed? _____

        Can the subject look at the ceiling and stand on one foot, with eyes open? _____

        With eyes closed? _____

Copyright © 2015, 2011, 2008 Pearson Education, Inc.

Summarize your observations and explain the differences observed. _____
_____
_____

6. Reaction times

Trial	Reaction Time					
	1	2	3	4	5	Average
Sight, one color						
Sight, choice of two colors						
Sight, choice of three colors						
Sight, choice of four colors						
Sound						
Word association	Stimulus word			Response word		Reaction time

How do you explain the differences obtained for the various reflex and reaction times?

_____

6a. Reaction times, Biopac version.

Male _____ or Female _____

Student's Name	Reaction Times (ms)			
	Pseudo Random Trial 1 data		Fixed Interval Trial 1 data	
	Stimulus 1	Stimulus 5	Stimulus 1	Stimulus 5
Means				

Copyright © 2015, 2011, 2008 Pearson Education, Inc.

Which trial had the lowest reaction time? _____

Explain what differences you observed. _____

_____

Student's Name	Reaction Times (ms)			
	Pseudo Random Mean		Fixed Interval Mean	
	Trial 1	Trial 2	Trial 1	Trial 2
1				
2				
3				
4				
5				
6				
7				
8				
9				
10				
Mean				

Describe the changes observed in the mean reaction time for stimulus 1 and 5 for each trial._____

_____

Explain your observations. _____

_____

Males

Class Data Student Names	Reaction Times (ms)			
	Pseudo Random Mean		Fixed Interval Mean	
	Trial 1	Trial 2	Trial 1	Trial 2
1				
2				
3				
4				
5				
Class Mean				
Standard Deviation				

Copyright © 2015, 2011, 2008 Pearson Education, Inc.

Females

Class Data Student Names	Reaction Times (ms)			
	Pseudo Random Mean		Fixed Interval Mean	
	Trial 1	Trial 2	Trial 1	Trial 2
1				
2				
3				
4				
5				
Class Mean				
Standard Deviation				

Describe any differences in the means for males and females. _____

_____

Use a scientific calculator to calculate the standard deviation (SD) for each group.

SD: Male _____ SD: Female _____

Does the mean reaction time for the male or female subjects lie within one or two standard deviations of the other sex? _____

How do you interpret your results for the differences between reaction times between males and females when you consider the standard deviations for each trial?_____

_____

APPLY WHAT YOU KNOW

1. Because all action potentials are all-or-none, how do we distinguish between stimuli of different intensities?

2. Because all stimuli are ultimately transformed into action potentials before being delivered to specific areas in the brain, how do we recognize the difference between, for instance, a visual stimulus and a chemical stimulus?

3. What would you perceive if the optic nerve were mechanically stimulated? Explain your answer.

Copyright © 2015, 2011, 2008 Pearson Education, Inc.

Sensory Physiology I: Cutaneous, Hearing

OBJECTIVES

After completing this exercise, you should be able to

1. Examine and explain the functioning of cutaneous receptors.
2. Examine and explain the functioning of hearing receptors.
3. Differentiate between conductive and sensori-neural deafness.

Sensory Receptors

Our knowledge of changes in our environment depends on the sensory nervous system and its receptors. Classically, we speak of five senses: sight, hearing, touch, taste, and smell. In this laboratory exercise, you will study the mechanism of action of several **exteroceptors**—receptors that receive stimuli from outside the body. Exteroceptors are usually located on the surface of the body, in contrast to **interoceptors**, which are located deep within muscles, tendons, and other structures and which detect changes within the body.

▶ ACTIVITY 7.1
Cutaneous Receptors

Materials (per group)

- ☐ 1 compass or aesthesiometer
- ☐ 1 sharpened pencil
- ☐ A selection of pennies, dimes, and nickels (at least 3 of each)
- ☐ 1 2-in.-square piece of cardboard
- ☐ 1 brass weight set ranging from 1 g to 200 g
- ☐ 3 1,000-ml beakers filled with 500 ml each of ice water, 25°C water, and 45°C water

1. Tactile Distribution: Two-Point Sensibility

Receptors for touch vary in their density of distribution over various areas of the body surface. Areas of the body with many touch receptors,

such as the fingers, have a finer sense of feel—tactile discrimination.

a. Have the subject sit with the eyes closed. Use a compass that has blunted tips, a caliper, or an aesthesiometer to apply tactile stimuli to the subject's skin. Start with the two points close together and then increase the distance between them until the subject feels two distinct points. Be sure the two points are applied simultaneously each time.

b. Apply one and two points randomly to test whether the two-point threshold has been reached or whether the subject's imagination is operating overtime. **Record in the Laboratory Report the point distance in millimeters for the following body areas:**

   Back of neck 1.7cm , 4.8cm
   Fingertip 1cm , 1.8 cm
   Forearm (palmar surface) 1.5cm , 5cm
   ~~Tip of nose~~
   Palm of hand 2.1cm , 2.1cm
   ~~Tongue~~
   Upper arm 3.5cm , 7.4 cm

2. Tactile Localization

a. Have the subject close the eyes. Touch the subject's skin with a pointed pencil to leave an indentation. Remove the pencil. Then have the subject try to touch this exact spot using another pencil. Measure the error of location in millimeters.

b. Repeat, using the same stimulation point. Does the subject's localization improve the second time? **Record the localization error distance (two trials) for the following body regions:**

   Palm

   Fingertip

   Forearm

   Lips

3. Adaptation of Touch Receptors

a. Close your eyes and have your partner place a small coin on the inside of your forearm.

Copyright © 2015, 2011, 2008 Pearson Education, Inc.

Determine how long (in seconds) the initial pressure sensation persists. Repeat the experiment at a different forearm location, and when the sensation disappears, have your partner add two more coins of the same size. Does the pressure sensation return, and, if so, how long does it last with the added coins? What receptors are functioning here, and why is the sense of pressure soon lost?

b. Another illustration of sensory adaptation is provided by the touch receptors around the root of each hair. Using a pencil point, move one hair as slowly as possible until it springs away from the pencil. Is the touch sensation greater when the hair is slowly bent or when it springs back? These hair receptors have one of the fastest adaptation times to stimulation of all the receptors. **Enter your observations in the Laboratory Report and answer the associated questions.**

4. Weber's Law

Weber's law states that the size of the just noticeable difference, ΔI, is a constant proportion of the original stimulus, I. It can be demonstrated using the following directions.

a. Blindfold the subject and have him or her place one hand, palm up, on a table. Place a 2-in.-square piece of cardboard on the distal phalanges of the index and middle fingers. Place a 10-g weight on the cardboard. After the subject feels the weight, remove the cardboard and weight, add additional weights of 1 to 5 g, and replace the cardboard and weights on the fingers.

b. Repeat this procedure until the subject reports a perceptible sensation of increased weight compared with the initial weight of 10 g. Record the increment (added weight) required to produce the sensation of added weight in the Laboratory Report. This is called the **just noticeable difference (JND)** or **intensity difference** (ΔI). The ratio of intensity difference to the initial weight intensity is called **Weber's fraction** ($\Delta I/I$). **Note:** The cardboard and weights should be lifted from the fingers while additional weights are being added and then all placed back on the fingers at once.

c. Repeat the experiment, starting with an initial weight of 50 g, increasing to 100 g, and finally ending with 200 g. Record the weight increments required for the subject to perceive a JND in weight for each of these initial weights in the Laboratory Report. You should occasionally use the same weight in consecutive tests

to validate your observations. In each trial, what is the ratio of the JND to the initial weight (Weber's fraction)? How does this ratio compare between the various initial weights? Note that Weber's law holds only for weights in the medium range. **Enter your observations in the table in the Laboratory Report.**

5. Temperature Receptors: Adaptation and Negative Afterimage

a. Prepare three 1,000-ml beakers half full of ice water (0–5°C), water at room temperature (25°C), and water at 45°C, respectively.

b. Place your left hand in the ice water and your right in the warm water. What happens to your sensation of cold or warmth in each hand after 2 min? Which hand seems to adapt fastest?

c. Now rapidly place both hands in the room-temperature water. What are your sensations in each hand? This experiment illustrates that the sensations of heat and cold are not absolute but depend on how rapidly the skin gains or loses heat and on the magnitude and direction of the temperature gradient. **Enter your observations in the Laboratory Report.**

6. Referred Pain

Referred pain is the strange phenomenon of perception of pain in one area of the body when another area is actually receiving the painful stimulus. We say the pain is referred to the other, more remote area. This phenomenon explains why a heart attack might cause a patient to feel pain in the neck, shoulder, or arm rather than in the chest. Place your elbow in ice water and, over a period of time, note any changes in the location of sensation perceived. Does it change location? If so, where is the referred pain felt? The ulnar nerve, which supplies the ring finger, little finger, and inner side of the hand, passes over the elbow joint and serves as the mediator for this referred pain sensation. You might have experienced other examples of referred pain such as pain in the forehead after swallowing ice cream, a phenomenon commonly known as a brain freeze or an ice-cream headache. **Enter your observations in the Laboratory Report.** ◀

● STOP AND THINK

The trigeminal nerve divides into the ophthalmic, maxillary, and mandibular branches. Among other functions, the maxillary branch provides sensory inputs from the palate and the pharynx to the brain, whereas the ophthalmic branch does the same for the forehead. How would these anatomical aspects help explain a brain freeze?

Copyright © 2015, 2011, 2008 Pearson Education, Inc.

▶ ACTIVITY 7.2

Hearing

Materials (per group)

☐ 1 stethoscope

☐ 1 ticking watch or a few coins

☐ 1 512-Hz tuning fork

☐ Cotton balls

☐ 1 audiometer

1. Watch Tick Test for Auditory Acuity

This activity should be performed in a quiet room. Although an old-fashioned ticking watch is used as an example, quietly tapping two coins together or any other source of consistent sound can be substituted for the watch.

a. Have the subject plug one ear with cotton and close his or her eyes.

b. Hold a watch in line with the subject's open meatus. Gradually move the watch away from the ear until the subject just loses the ability to hear the ticking. Measure the distance (in centimeters) between watch and ear and record.

c. Move the watch farther away and then begin moving it closer to the subject's ear until the subject first hears the ticking. Is this distance the same as when the watch was moving away?

d. Test the other ear in the same manner. Is the acuity the same for both ears? This is not a fair test for an elderly person because high tones are lost first in old age. **Enter your observations in the Laboratory Report.**

2. Localization of Sound

a. With the subject seated and blindfolded, bring a watch within hearing range from several different angles around the subject's head. (The clicking of two coins together can be used in place of the watch in this test.)

b. Ask the subject to point to the direction from which he or she hears the sound. Is the subject's judgment better in the median plane or at the side of the head? In the median plane, is the subject more accurate with sound above the head or in front of it?

EXPLAIN THIS!

The ability to localize sound is a consistent behavior observed in humans. How do the ears allow us to localize sound and what brain structures are involved in helping us to accurately turn our heads toward the source of the sound?

---

3. Auditory Adaptation

a. Place a stethoscope in the subject's ears. Place a vibrating tuning fork near the diaphragm of the stethoscope so that the sound seems equally loud to both ears. Remove the tuning fork and wait a minute or two.

b. Pinch the tube to one ear to occlude the tube and place the vibrating tuning fork in its former position near the diaphragm. When the sound becomes nearly inaudible to the open ear, open the pinched tube. Is this ear also adapted to the sound? **Explain your observations in the Laboratory Report.**

4. Tuning Fork Tests   ·hearing evenly

These two simple hearing tests are used to distinguish between conduction and sensorineural deafness. In conduction deafness, transmission of sound waves through the middle ear to the oval window is impaired. In sensorineural deafness, one transmission of nerve impulses from the cochlea side to the auditory cortex is impaired. A 512-Hz tun-mu ing fork is preferred for these tests because most people have difficulty telling whether the vibrations are felt or heard when forks of lower frequency are used. The fork should be struck on a soft surface such as the heel of the hand because striking a hard surface produces overtones.

Weber Test This test should be performed in a room with a normal noise level (not quiet). Strike the tuning fork and place the tip of the handle against the middle of the subject's forehead. If both ears are normal, the tone will be equally loud in both ears, and the sound will be localized as coming from midline position. If an individual has conduction deafness, the sound will be heard louder in the deaf ear than in the normal ear because, in the normal ear, the sound will be partially masked by environmental noise to which the defective ear is less sensitive. Also, in the normal ear, the sound is damped, or softened, by the tensor tympani and stapedius muscles, which prevent the full amplitude of vibration of the auditory ossicles. This attenuation reflex is less effective or is absent in conduction deafness.

If an individual has a defect in the auditory nerve or cochlear apparatus (sensorineural deafness), the sound will be heard better in the normal ear than in the deaf ear because neural activity is essential for hearing.

Conduction deafness can be simulated by plugging one ear with cotton.

Rinne Test This test compares air conduction of sound with bone conduction of sound. It should be performed in a quiet room. Locate

Copyright © 2015, 2011, 2008 Pearson Education, Inc.

the mastoid process of the temporal bone behind the ear of the subject. Strike the tuning fork and place the handle against the mastoid at the level of the upper portion of the ear canal. As soon as the sound is no longer audible through the bone, hold the vibrating prongs of the tuning fork about 1 in. from the ear, and the subject should again be able to hear the sound. A person who has normal hearing will hear the sound several seconds longer through air conduction than through bone conduction because the threshold for air conduction is lower. A person who has conduction deafness will hear the sound as long as or longer through bone conduction than through air conduction. The person with nerve deafness will hear the sound longer through air conduction than through bone conduction but usually requires a louder sound to hear at all.

A concise guide for interpreting the Weber and Rinne tests is given in the table on the following page.

5. Audiometry

Audiogram A pure-tone audiometer is an instrument for measuring hearing acuity. It consists of an earphone connected to an electronic oscillator capable of producing pure tones in the range of 250 to 8,000 Hz. This range is narrower than the hearing range for normal humans (20 to 20,000 Hz). Dogs and very young children often can hear frequencies as high as 40,000 Hz.

In this test, the subject listens to several tones in the range of 250 to 8,000 Hz, usually at octave or half-octave intervals. The subject's threshold of perceiving each tone is determined and recorded as decibels (dB) of loudness needed to just hear each tone. If the loudness of a tone must be increased to 20 dB above the normal tone level, the subject is said to have a **hearing loss** of 20 dB for that particular tone. The amount of hearing loss is plotted on the **audiogram** below the normal hearing line at zero.

Test the auditory acuity of the subject, starting with high frequencies (8,000 Hz). Determine the number of decibels required for the subject to hear each tone. Decrease the frequency in octave intervals (4,000 Hz, 2,000 Hz, 1,000 Hz, 500 Hz, and so on). Plot the audiogram of the subject's hearing loss on the chart provided in the Laboratory Report. The audiometric zero (0 dB) (normal) is based on extensive tests of young persons with normal hearing. Here, audiometric zero (0 dB), which is usually considered the average normal hearing threshold level, is not actually the point at which impairment begins. According

to standards published by the American Speech-Language-Hearing Association (ASHA), hearing loss (HL) levels are expressed in decibels based on average thresholds for frequencies in the 500 to 4,000 Hz range.

Normal hearing loss	0–20 dB HL
Mild hearing loss	20–40 dB HL
Moderate hearing loss	40–60 dB HL
Severe hearing loss	60–80 dB HL
Profound hearing loss	80 dB or higher HL

Additionally, qualitative factors such as bilateral versus unilateral loss, frequency-specific hearing loss, and stable versus fluctuating hearing loss are, among other symptoms, evaluated during a hearing examination.

How would an audiogram for age-related deafness (presbycusis) look? **Enter your data in the table in the Laboratory Report.**

CLINICAL APPLICATION

Hearing Aids

The type of hearing aid used to compensate for hearing loss depends upon the cause of deafness, either conductive, sensorineural, or mixed hearing loss. In the case of conductive loss, a patient will use a hearing aid that consists of a microphone, an amplifier, a processor that shapes the sound output to best match the hearing loss, and a receiver/speaker that directs the processed sound to the ear canal. If one is diagnosed with sensorineural hearing loss, a cochlear implant is used to directly stimulate the auditory nerve in the inner ear. The external parts of a cochlear implant act very much like a hearing aid in that they pick up sound and send it to a speech processor that digitizes the signals and sends them to an implanted receiver under the skin. The coded signals are ultimately delivered to a series of electrodes that directly stimulate the auditory nerve in the cochlea. In the case of mixed hearing loss, an audiologist will typically take care of the conductive loss first.

Evaluation of Hearing Impairment (Shorter Method) In 1959, the American Academy of Ophthalmology and Otolaryngology issued a statement describing a simpler method for the evaluation of hearing impairment. This method recognizes that hearing impairment should be evaluated in terms of ability to hear everyday speech under everyday conditions and that everyday speech does not encompass the entire audiometric range but, primarily, the frequencies

Copyright © 2015, 2011, 2008 Pearson Education, Inc.

from 500 to 2,000 Hz. Therefore, for a practical evaluation of everyday speech impairment, the academy recommends determining the simple average of hearing levels at frequencies of 500,

1,000, and 2,000 Hz, using a pure-tone audiometer for testing. **Use your audiogram to enter data in the table in your Laboratory Report.** Note that hearing loss is averaged for each ear. ◄

Guide for Interpreting Weber and Rinne Tests		
Condition	Finding	
	Weber Test	Rinne Test
No hearing loss	No lateralization	Sound heard longer through air conduction
Conduction deafness	Lateralization to the deaf ear	Sound heard as long as or longer through bone conduction
Sensorineural deafness	Lateralization to the normal ear	Sound heard longer through air conduction

Right: 21 500   19
9 1000   14
12 2000   11

37 500   19
21 1000   9
10 2000   10

Copyright © 2015, 2011, 2008 Pearson Education, Inc.

Copyright © 2016 L. L. 2000 Pearson Education, Inc.

Sensory Physiology II: Vision

OBJECTIVES

After completing this exercise, you should be able to

1. Examine and test the functions of the eye.
2. Understand and experimentally demonstrate the optics of vision.
3. Dissect the functional anatomy of the eye.
4. Use the ophthalmoscope to examine the retina.

Functions of the Eye

Organs of vision range in complexity from simple eye spots, which can detect the presence or absence of light, to the more complex eyes seen in cephalopod mollusks, birds, and mammals. All complex eyes share the following common functional features: an ability to regulate the amount of light that enters the eye, an ability to focus light on photoreceptors, and the ability to interpret the nerve impulses generated by the photoreceptors. In this lab, you examine the anatomy and physiology of the eye to understand how it performs these functions.

► ACTIVITY 8.1

Vision and the Functions of the Eye

Materials (per group)

☐ 1 flashlight

1. Accommodation Reflexes
Accommodation reflexes are often studied under the topic of reflexes because they represent a programmed response to a stimulus.

Observe the pupil of your partner's eye under normal lighting. Shine a light into the eye and notice the constriction of the pupil. The purpose of this pupillary reflex is quite obvious. Now have your partner focus on an object across the room and observe the size of the pupil. Then have your partner focus on an object 6 in. from the eye and observe the size of the pupil. *smaller* **Enter your observations and answer the questions in the Laboratory Report.**

across: bigger

2. Near Point of Accommodation

Materials (per group)

☐ 1 meter ruler

☐ 13 x 5 in. unlined card

To produce a sharp image on the retina, the lens of the eye must be able to change its focusing power for viewing objects at different distances. When viewing a near object, the lens becomes more spherical than when viewing a distant object. As we age, the lens becomes less elastic and, therefore, less able to form the spherical shape needed for near vision. To compensate for this loss of accommodation power with age (presbyopia), persons over the age of 45 begin wearing bifocal or trifocal glasses to give them better near vision. Determination of the near point for the eye gives us a measure of the elasticity of the lens and its accommodation power.

The *near point* is the closest distance at which one can see an object in sharp focus. Print a small letter, 5 mm high, on a white card. With one hand, hold a meter stick directed forward from the bridge of your nose. Hold the card at a distance on the meter stick and close one eye. Move the card toward your eye until the letter becomes blurred, then move it away until the letter is a clear image. Measure the distance from the card to your eye. Repeat the test for the other eye and record your results.

Compare your near point measurements with the following normal values:

110 10
92 11.3

τ.86 ~ 6.5 L
~ 100 ~ 9 R

Age (Years)	Near Point (cm)
10	9
20	10
30	13
40	18
50	53
60	83
70	100

Enter your observations and answer the questions in the Laboratory Report.

Copyright © 2015, 2011, 2008 Pearson Education, Inc.

3. Binocular Vision and Space Perception

Materials (per group)

☐ 1 large die

These simple experiments illustrate the advantage of binocular vision in providing depth of field and space perception.

Hold a cube 3 in. in front of your nose and focus on it. Close one eye, open it, and close the other. Repeat this procedure while looking at a picture in a stereoscope. Hold your finger about 6 in. in front of your nose. Look at a distant object with both eyes open. The finger will double. Now cover the right eye. **Enter your observations and answer the questions in the Laboratory Report.**

4. Blind Spot

The blind spot is an area on the retina where the optic nerve and blood vessels enter and leave, hence where there are no rods or cones for visual reception.

Close your left eye and focus your right eye on the image of the plus symbol (+) on the left of Figure 8.1. Hold the paper about 15 in. from the eye and slowly bring it closer. Soon the image of the dot will disappear. At this distance, the dot image is being focused on the blind spot, the area of the right eye where there are no rods or cones to perceive it. **Enter your observations and answer the questions in the Laboratory Report.** *25.5cm*

5. Visual Acuity and Astigmatism

Materials (per group) *24.3cm*

☐ 1 Snellen eye chart

Visual acuity is the power to discriminate details. You have tested your visual acuity many times using the **Snellen test** letters, but you might not have understood exactly what the score means. The basis for the Snellen test is that letters of a certain size should be seen clearly at a specific distance by eyes that have normal acuity. For example, line 1 of the test should be read easily at 200 ft, and line 8 at 20 ft (Figure 8.2).

A person's visual acuity is stated as $V = d/D$, in which d is the distance at which the person can read the letters and D is the distance at

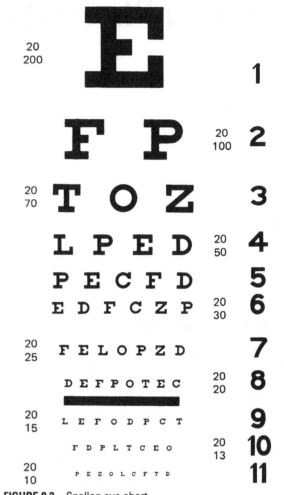

FIGURE 8.2   Snellen eye chart.

which a normal eye can read the letters. Stand *20/20* 20 ft from the full-sized Snellen chart on the wall (not the reduced chart in the manual). *20/20* Cover one eye and attempt to read line 8. If you can read it, your eye is normal and is rated as 20/20. If you can read line 9 at 20 feet, your visual acuity is above average, rated at 20/15. If you can't read line 8 but can read line 7, your visual acuity is below average, rated as 20/25. Repeat the test for the other eye. **Enter your observations and answer the questions in the Laboratory Report.**

Astigmatism is a condition in which there is an uneven curvature of the surface of the lens or cornea. This causes a greater bending of light rays as they pass through one axis of the lens or cornea than when light passes through another axis. This causes the image viewed to be blurred in one axis and sharp in the other axes.

Remove any corrective lenses, cover the left eye, and look at the center of the astigmatism test chart (Figure 8.3). If some of the lines appear blurred or lighter, this indicates that astigmatism

FIGURE 8.1   Blind spot demonstration.

Copyright © 2015, 2011, 2008 Pearson Education, Inc.

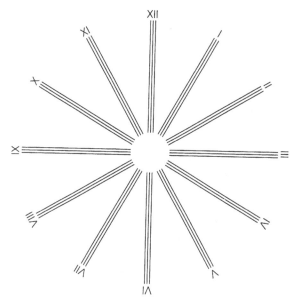

FIGURE 8.3 Astigmatism test chart.

is present. If all lines are equally sharp and black, no astigmatism exists. Repeat the test with the right eye. If you wear corrective lenses, repeat the tests with them on. **Enter your observations in the Laboratory Report.**

6. Negative Afterimages: Complementary Colors

[handwritten: QU: black, yellow green]

Materials (per group)

[handwritten: normal red blue white]

- ☐ 1 candle
- ☐ Matches or lighter
- ☐ 1 (each) red, green, and yellow 4 × 6 in. cards
- ☐ 1 unlined white sheet of paper

Look at a very bright scene (such as a burning candle) for about 30 sec and then shift your gaze quickly to a blank white surface. What sort of image do you perceive? A negative afterimage should be apparent in which the light areas of the original scene appear dark and the dark areas appear light. This perception is explained in the following manner: While you are viewing the bright scene, cones receiving light from the bright areas become **light adapted**, or bleached of their visual pigment, and when you shift your gaze to the white paper, these cones cannot initiate impulses. Thus, this area of the scene now appears dark. The unadapted cones are still sensitive and are stimulated by the white background.

Color vision is initiated by the activation of cone cells in the retina that are sensitive to red, green, and blue wavelengths of light. Other colors are perceived by differential stimulation of these three types of cone cells. The color white is perceived

when all the visible light rays are combined and the three types of cones are equally stimulated. White is also perceived when two particular colors are mixed, so-called complementary colors. The relationship between the red, green, and blue colors and the complementary colors that produces a white sensation is clearly portrayed in the physiologist's color wheel, shown in Figure 8.4.

Stare at a bright red color for a minute or so and then shift your gaze to a blank white sheet. What color is the afterimage? Repeat this procedure using a green color and then a yellow color. Record the color of the afterimage seen in each case. **Enter your observations and answer the questions in the Laboratory Report.**

7. Tests for Color Blindness

Materials (per group)

- ☐ 1 Ishihara test chart book
- ☐ 1 set of Holmgren's yarn

Color blindness is an abnormality that is transferred genetically, resulting from the lack of a particular gene in the X chromosome. The most common type is red–green color blindness in which the person lacks either the red or the green cones in the retina. If the red cones are lacking, red wavelengths of light stimulate primarily green cones and the person perceives red as green. If the green cones are missing, the person will perceive only red colors in these wavelengths of light. **Ishihara test** These test charts are probably the most widely used in the testing of color blindness. Each chart contains different-colored dots arranged so that the person with normal color

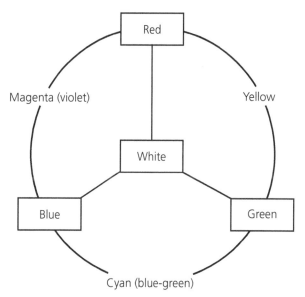

FIGURE 8.4 Physiologist's color wheel.

Copyright © 2015, 2011, 2008 Pearson Education, Inc.

vision reads one number and the color-blind person perceives a different number.

Hold the charts about 30 in. away from you and in a good reading light (avoid intense light because the colors will fade). Read each plate. How many color-blind persons are detected in the class? **Holmgren's test** This test consists of several skeins of colored yarn that must be matched accurately with three sample colors, A, B, and C. The odd-numbered skeins are the same color as the A, B, and C colors, but the even-numbered skeins are "confusion" skeins and should not be mistaken for the true color-matched skeins. A color-blind person will match some of the "confusion" skeins with the A, B, or C colors because they appear to be the same. **Answer the questions in the Laboratory Report.**

EXPLAIN THIS!

We have three types of cones, those sensitive to the colors red, green, and blue. How do you explain how we see more than three colors? Use specific colors like purple and brown to explain your answer.

---

8. The Optics of Vision

Materials (per group)

☐ 1 Cenco or Ingersoll eye model with associated lenses

The primary task of most of the structures in the eye (except the retina) is simply to aid in focusing light rays on the retina. The function of various refractive parts of the eye can be demonstrated by the Ingersoll or Cenco eye models. These models consist of a water-filled black tank with chambers representing the aqueous and vitreous humors. In the anterior end, a curved piece of glass performs the function of the cornea. Behind and in front of this "cornea" are slots for holding various "eye" lenses (and irises) and corrective lenses. At the posterior end of the model is a movable metal screen that serves as the retina, on which the image is focused. It can be placed in three positions to simulate normal, farsighted, and nearsighted vision. In most models, a white spot in the middle of the retina represents the fovea, and a dark spot to one side represents the blind spot. See Figure 8.5.

A complete set of lenses includes the following:

1. Double convex, +7.00 diopters, spherical convergent
2. Double convex, +20.00 diopters, spherical convergent
3. Double convex, +2.00 diopters, spherical convergent
4. Double concave, −1.75 diopters, spherical divergent
5. Concave cylindrical, −5.50 diopters, divergent
6. Convex cylindrical, +1.75 diopters, convergent
7. Iris diaphragm disk, 13-mm-diameter aperture

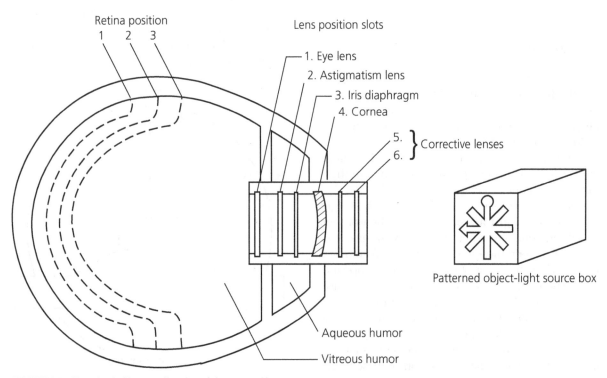

FIGURE 8.5 Top view of Ingersoll eye model.

Copyright © 2015, 2011, 2008 Pearson Education, Inc.

The refractive power of a lens is measured in **diopters:**

$$\text{Diopters} = \frac{1}{\text{Focal length (m)}}$$

Thus, a lens that can focus light at 17 mm (0.017 m) has a power of 1/0.017 m = 59 diopters. This is about the average refractive power of the normal adult eyeball (the length of the eyeball is about 17 mm). The larger the diopter rating, the stronger the refractive power of the lens.

Normal distance vision Place the eye model on a table facing a bright object (such as a window) 4 to 5 m away. Fill the tank to within 2 cm of the top with clear water. Add a few drops of eosin to trace the path of the light rays from the lens to the retina.

Place the retina in the normal position (slot 2) and the +7.00 lens in slot 1.

Normal near vision and accommodation Face the eye model toward the patterned object–light source box, placed at a distance of exactly 33 cm from the cornea (lens position slot 4). Is the image in focus on the retina? Replace the +7.00 lens with the +20.00 lens in position 1. Compare the shapes of these two lenses. The image should now be in focus; if it is not, move the light source a little. This comparison illustrates how the eye lens thickens to accommodate for viewing near objects. **Enter your observations and answer the questions in the Laboratory Report.**

Leave the +20.00 lens in position 1 for experiments 3 through 8.

Effect of pupil size Move the light source slightly until the image on the retina is a little out of focus. Insert the iris diaphragm disk (13-mm-diameter hole) in position 3 behind the cornea. **Enter your observations and answer the questions in the Laboratory Report.**

Hypermetropia (farsightedness) Place the object light source 33 cm from the cornea as before and move the retina to position 3. Note the character of the retinal image and its size. Correct this defect by moving the object closer to or farther from the cornea. Which works? Have you ever seen anyone use this technique for correcting farsightedness? **Enter your observations and answer the questions in the Laboratory Report.**

Now return the object to the 33-cm position and try to correct the image using a corrective lens in front of the cornea. Try a +2.00 or –1.75 lens. Which corrects it—a converting or diverging lens? **Enter your observations and answer the questions in the Laboratory Report.**

Myopia (nearsightedness) Remove the corrective lens and place the retina in position 1. Note the character and size of the image. Try to focus the image by moving the object. Return the object to the 33-cm position and try to correct the defect by using a +2.00 or –1.75 lens in front of the cornea. **Enter your observations and answer the questions in the Laboratory Report.**

Astigmatism Remove the correcting lens and leave the object lamp at 33 cm from the cornea. Place the retina in the normal (2) position. Now insert the concave, cylindrical –5.50 lens in position 2 behind the cornea. Note that now the image is blurred along certain lines of the object. Such astigmatism in the human eye is produced by an irregular curvature of the cornea. The –5.50 lens allows you to simulate such a refractive defect.

Now rotate the cylindrical lens and note how different lines of the object become blurred while others remain sharp. Correct for this astigmatism by placing the convex, cylindrical +1.75 lens in front of the cornea and rotating it to focus the image. Compare the relative directions of the cylindrical axis in each of these lenses when the correction is made. A similar corrective lens is used in your own eyeglasses to correct your astigmatism. **Enter your observations and answer the questions in the Laboratory Report.**

Compound defects Produce a compound defect in the model eye of astigmatism by adding either myopia or hypermetropia. Try to correct this compound defect by using the proper combination of lenses. Such compound defects are common in human vision, and the corrective lenses are combined in one eyeglass lens or contact lens. **Enter your observations and answer the questions in the Laboratory Report.**

Action of a magnifier Place the retina in the normal position, the object at 33 cm, and the +20.00 lens in position 1. Use the +7.00 lens as a magnifying lens in front of the cornea. Try to focus the image on the retina by moving the object lamp either closer to or farther from the cornea. **Enter your observations and answer the questions in the Laboratory Report.**

Lens removal In cataract disease, the lens of the eye becomes opaque and vision diminishes to near blindness. Cataracts are corrected by removing the lens and substituting a corrective eyeglass lens to take the place of the eye lens. Stimulate this correction by removing the +20.00 lens. Now place the +7.00 lens in front of the cornea and describe the image appearance. Bring the object closer and observe the image. These adjustments give you some idea of the corrections and relative degree of restored vision for a person who has cataract disease. **Enter your observations and answer the questions in the Laboratory Report.**

Copyright © 2015, 2011, 2008 Pearson Education, Inc.

handwritten note at top:
-1.75 → moves w/ u and is clear
+2.00 → reversed; blurry
goes opposite

Examination of eyeglass lenses Look through the +2.00 lens and move it from side to side. Note whether the object viewed moves in the same direction or in the opposite direction as the lens is moved. Repeat this, using the −1.75 lens. Examine your own eyeglasses, or those of your partner, and determine whether they are concave or convex and what defect they are meant to correct. Do the two eyeglasses have the same kind of correction? Rotate the eyeglasses. Is there any correction for astigmatism? Compare it with the rotation of the −1.75 convex cylindrical lens. **Enter your observations and answer the questions in the Laboratory Report.**

At the end of the experiment, remove all lenses from the model, clean them with lens paper, and replace them in the lens case.

9. Visual Fields of the Eye: Perimetry

Materials (per group)

☐ 1 perimeter

The field of vision is the entire area seen by the eye when fixed in one position. Charting the visual field is useful in localizing brain lesions or determining blindness in various retinal areas. The instrument used to map the field of vision is called a **perimeter**. It consists of a black metal semicircle graduated in degrees from 0° at the center to 90° at the edges. The subject focuses on a white dot in the center of the perimeter during the test. The operator moves a pointer along the perimeter until it just enters the subject's field of vision; then the degrees from center are read and plotted on a perimetry chart. By repeating this in different planes and for different colors, a map of visual fields for rod and cone vision is obtained. Modern perimeter machines use computers and monitors to evaluate a patient's field of vision interactively.

a. Have the subject sit at a table with the chin on the chin rest. Adjust the perimeter so the white dot is level with the eye and the ends of the perimeter semicircle are aligned with the eye. The untested eye is covered or closed. The perimeter should be in the horizontal or 0° position.

b. While the subject stares fixedly at the white dot in the center, the operator (seated behind the perimeter) slowly pushes the carrier with its white disk along the perimeter from the outer edge toward the center. When the dot is just visible, the subject informs the operator, who then records the position on the perimeter. Usually, three tests are run at each perimeter angle, and the average is recorded on the perimetry chart. Note that the readings taken from the right side

of the semicircle are plotted on the left side of the chart and vice versa. Repeat the procedure for the opposite side of the perimeter.

c. Replace the white disk with a colored disk (red, blue, or green) and repeat the perimetry measurements. Do this for all three colors. The subject must correctly identify the color—not just the disk movement—before the degree of angle is recorded. Randomly mix the colors so that the subject cannot memorize any sequence.

d. Repeat the procedures using the same eye but with the perimeter rotated 30°, 60°, 90°, 120°, and 150° from the original position. Do this with the white disk and with the three colored disks. Plot the visual fields for these on the same chart.

How do the fields compare in size? What might cause a decrease in the size of the visual field? If one of the members of the class knows that he or she has an altered visual field, it would be interesting to plot the field and compare it with a normal field.

A perimeter chart of the right eye with the fields of vision for green, red, blue, and white is shown in Figure 8.6. Note that the field for green is the most limited, with red, blue, and white having more expanded visual fields. How do your fields compare with those shown for a normal eye? **Enter your observations and answer the questions in the Laboratory Report.** ◀

Anatomy of the Eye

Perhaps more so than for the other organs of the body, the student needs a thorough knowledge of the anatomy of the eye to understand how it works. Although some knowledge can be obtained by studying diagrams and models, there is no adequate substitute for actual dissection of an eye. Cow or sheep eyes will be used because they are large and available. Fresh or frozen eyes are superior to those preserved in formalin because the lens retains its clarity and elasticity.

▶ ACTIVITY 8.2
Anatomy of the Eye

Materials (per student)

☐ 1 cow or sheep eye
☐ 1 (each) scissors and forceps

1. External Features
Examine your laboratory sample. Notice the stump of the **optic nerve** at the posterior of the eye. Because this nerve is directed toward the midline

Copyright © 2015, 2011, 2008 Pearson Education, Inc.

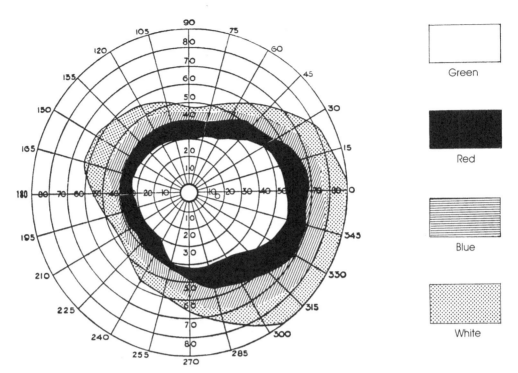

FIGURE 8.6   Perimeter chart of right eye.

of the body in its natural position, you can use it as a guide in orienting the eye. The front surface of the eye has a clear window, the **cornea**, through which light enters. The cornea is covered by a thin transparent epithelium, the **conjunctiva**, which merges on all sides with the connective tissues that aid in holding the eye in place. These tissues are also called conjunctiva, although in the strict sense, the term applies only to the epithelium. The tough white outer wall of the eye, composed of fibrous connective tissue, is the **sclera**. Attached at various points on the sclera are the **extrinsic** or **extraocular muscles**, which move the eye as a whole. There are three pairs of these muscles, and their names describe their positions on the eye: the **superior** and **inferior recti** (singular, rectus) on the top and bottom, respectively; the **lateral** and **medial recti**; and the **superior** and **inferior obliques**. These last two muscles are attached to the eye near the superior and inferior recti, respectively, but the direction of their fibers is toward the medial wall of the eye socket. Look into the eye through the cornea and find the **iris**, a membranous curtain that divides the eye into anterior and posterior chambers, and the **pupil**, an opening in the iris that allows light to pass farther into the eye.

2. Internal Features

Using scissors, carefully cut the eye into anterior and posterior halves. Refer to Figure 8.7 while examining the internal structures. The semigelatinous substance behind the lens is called the **vitreous humor**. Examine the posterior half of the eye first. Note that the wall of the eye is actually made of three main layers: the **sclera**, a vascular and darkly pigmented layer called the **choroid**, and the dark, innermost layer, called the **retina**. The retina actually has three layers: the **sensory layer**, which contains the modified neurons (rods and cones) that are the light receptors; a layer of bipolar cells; and a layer of ganglion cells. Ganglion cell axons form the optic nerve. Examine the surface of the retina closely and find the area where the fibers of the optic nerve and numerous blood vessels enter and fan out over its surface. This area, called the **optic disk**, contains nerve fibers and blood vessels instead of rods and cones and, thus, results in a blind spot. Lateral to the optic disk, you might find a slight depression that has a yellowish color. This is the **macula lutea**. The central part of this area, called the **fovea centralis**, contains only cones and provides the best focus of the image.

Examine the anterior half of the eye from the inside. The most prominent structure is the **crystalline lens**. This structure is held in place by about 70 radially arranged **suspensory ligaments** that are attached marginally to **ciliary bodies**, which are actually continuations of the choroid. The ciliary bodies contain **ciliary muscles**, which adjust the tension of

Copyright © 2015, 2011, 2008 Pearson Education, Inc.

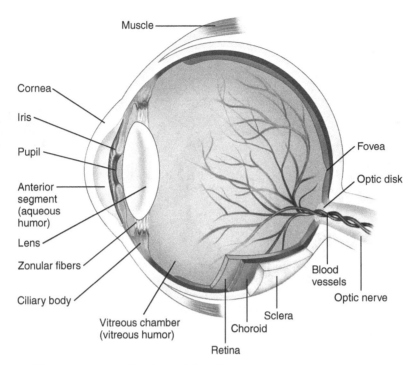

FIGURE 8.7 Anatomy of human eye as seen in a horizontal section.

Source: Stanfield, Cindy L.; Germann, William J., *Principles of Human Physiology*, 3rd Ed., © 2008, pg. 270. Reprinted and electronically reproduced by permission of Pearson Education, Inc., Upper Saddle River, New Jersey.

the suspensory ligaments and thus change the shape of the very elastic lens. Carefully remove the lens and note the delicate iris just in front of it. Can you determine that there are **radial** and **circular muscle fibers** in the iris? The space between the iris and the cornea is the **anterior chamber** and is ordinarily filled with a fluid-like substance called the **aqueous humor**. Grasp the edges of the lens with forceps and test its elasticity. Cut the lens open to determine its consistency. **Enter your observations and answer the questions in the Laboratory Report.** ◀

CLINICAL APPLICATION

The Cornea

The cornea is normally a transparent, curved tissue that covers the iris and the pupil and forms the anterior wall of the eye's anterior chamber. The cornea's curvature gives it refractive properties. These account for approximately two-thirds of the eye's total refractive power; unlike that of the lens, this refractive power is fixed.

Many different techniques are now used to reduce the need for corrective lenses by changing the shape of the cornea. Some of these techniques are laser-based surgical processes, and others use specialized contact lenses to reshape the cornea, both of which modify the refractive power of the cornea to allow better vision.

Ophthalmoscopy

Examination of the interior (fundus) of the eye with an **ophthalmoscope** can provide important information about the anatomy of the inner eyeball. It can also reveal some general health problems because the retina is one area of the body in which blood vessels can be examined without surgery. The view of the retina seen with an ophthalmoscope is called the **fundus oculi** (Figure 8.8). The main structures viewed in the fundus are the following:

Optic disk The spot on the retina where nerve fibers leave the eyeball and blood vessels enter and leave the eyeball. It contains no rod or cone cells and, thus, is a blind spot on the retina.

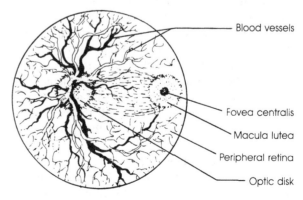

FIGURE 8.8 Fundus oculi.

Copyright © 2015, 2011, 2008 Pearson Education, Inc.

Macula lutea The yellow spot lateral to the optic disk, composed only of cone cells for **photopic vision** (bright light–color vision). The central pit of the macula, called the **fovea centralis**, contains a denser packing of cone cells, which provide greater visual acuity when light is focused on the fovea.

Peripheral retina The retinal areas peripheral to the macula that contain fewer cones and more rod cells and provide our **scotopic vision** (dim light–night vision) owing to the high sensitivity of the rods.

Blood vessels Retinal arteries and veins. These can alert a clinician to potential health problems. For instance, in diabetes, the peripheral vessels can be fewer and have small hemorrhage areas. In hypertension, vessels are of smaller diameter, and the veins are constricted where they cross the arteries. The larger, darker vessels are the veins.

The Ophthalmoscope

Before you examine your partner's eyes in the following activity, become acquainted with the operation of the ophthalmoscope (Figure 8.9). Light is beamed into the eye by means of a mirror inside the top of the instrument. A slit in the mirror allows the observer to see the interior of the subject's eye. The light intensity is varied by depressing the red lock button and rotating the **rheostat control** clockwise. Sharp focus on retinal structures is obtained by rotating the **lens selection disk** until the retina comes into focus.

The diopter rating of the lens is shown in the window, black numbers for convex and red numbers for concave lenses. If the eyes of the subject and examiner are both normal (**emmetropic**), no lens is needed, and an O appears in the window. The **aperture selection disk** on the front of the instrument changes the character of the light beam. A green spot is usually best because it is less irritating to the eye and displays the blood vessels more clearly.

The light beam should be directed toward the edge of the pupil, rather than through the center, to reduce the light reflection from the retina. It is important not to overexpose the retina with the strong light from the ophthalmoscope. *Limit the exposure time to 1 min* and then give the subject several minutes to rest before resuming the examination. Because you will not use drugs to dilate the pupil, you will not be able to examine the entire retina at once.

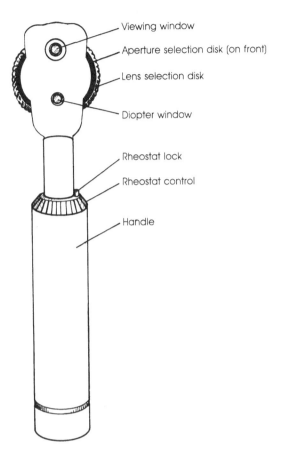

FIGURE 8.9   Ophthalmoscope.

▶ ACTIVITY 8.3

The Ophthalmoscope

Materials (per group)

☐ 1 ophthalmoscope

1. Sit facing the subject in a darkened room. Instruct the subject to focus on a distant object so that the pupil will be maximally dilated.
2. Use your right eye to examine the subject's right eye and your left eye to examine the left eye. Hold the ophthalmoscope with the index finger on the lens selection disk and your eye as close to the viewing window as possible. Retract the subject's upper eyelid with the thumb of your other hand.
3. Start viewing at a distance of about 12 in. Focus the beam on the pupil and examine the vitreous body and lens.
4. Keeping the pupil in focus, move close to the subject's eye (within 2 in.) and direct the beam near the edge of the pupil.
5. Rotate the lens selection disk to obtain a sharp image. Examine the fundus oculi for

Copyright © 2015, 2011, 2008 Pearson Education, Inc.

the optic disk, macula lutea, and blood vessels. Look for any irregularities on the retinal surface. To view the peripheral areas of the retina, ask the subject to look up, down, laterally, and medially. *Remember to limit your examination time to 1 min.*

6. Examine the subject's other eye; then change roles so the subject can examine your eyes. **Enter your observations and answer the questions in the Laboratory Report.** ◄

Copyright © 2015, 2011, 2008 Pearson Education, Inc.

Reproductive Physiology

⚠ CAUTION!

Because these experiments involve the use of living vertebrates, it is important that you obtain the appropriate permissions from your Institutional Animal Care and Use Committee (IACUC). While the instructions given are suitable to obtain the desired result, your IACUC may require a modification of procedures to meet your institution's needs.

OBJECTIVES

After completing this exercise, you should be able to

1. Describe the regulation of hormonal secretions and sexual function in the male and female.
2. Explain the physiology of the estrus cycle of the rat.
3. Identify the physiological basis of pregnancy tests.

Influence of Hormones on Reproduction

The interplay among the hypothalamus, pituitary, gonadotropic hormones, sex hormones, and sex organs is highly fascinating and quite complex. In these experiments, you examine the interrelationships of these hormones and organs, using rats that are castrated or ovariectomized, maintained for two weeks (some injected with replacement hormones), and then sacrificed to examine the effects on certain reproductive organs. You also have the opportunity to examine the five-day cycle of estrus, or heat, in the female rat.

It will be your responsibility to perform the surgical operations, care for the animals, give injections when needed, make the required dissections, and present your results to the rest of the class for discussion on the last day of this experiment.

You are urged to read as much as possible about reproductive physiology so that you can properly explain the results of the experiments. Although the experiments themselves are somewhat dogmatic, their presentation and explanation should open many areas for discussion and thinking. Before you begin, acquaint yourself with the reproductive system in the rat by studying Figures 9.1 and 9.2 and observing any available demonstration specimens.

General Directions for Aseptic Operations

These experiments on the endocrine system are performed on chronic animals, that is, animals that are kept alive for an indefinite period following the operation. Rats are quite resistant to infection, so successful operations can be performed with a minimum of aseptic techniques; however, it is strongly recommended that all instruments be autoclaved prior to use.

1. Instruments used in the operation should be sterilized in an autoclave.
2. It is best to work in teams of two during the operation, with one person acting as the anesthetist and the other as the surgeon. The anesthetist anesthetizes the rat, clips the hair from the operation site, and swabs the skin with alcohol and Betadine®. It is the anesthetist's job to watch the animal closely during the operation and regulate the depth of anesthesia (we want our patients to live). The surgeon scrubs his or her hands thoroughly using a Betadine surgical scrub before the operation, performs the operation, and closes the incision with sutures or wound clips.
3. Weigh the animal to determine the dosage of the anesthesia. Record the weight.
4. *All activity involving isoflurane should be performed under a ventilation hood.* Initially, anesthetize the animal by placing it in a desiccator containing a wad of cotton soaked with isoflurane. As soon as the rat becomes unconscious, remove it from the desiccator and use an isoflurane nose cone to maintain the depth of anesthesia. The nose cone is a large test tube containing a small isoflurane-soaked ball of cotton. Plug the test tube with dry cotton when it is not being used.
5. Inject the animal intraperitoneally with a ketamine-xylazine cocktail. Make the injection with the head of the rat tilted down, on either side of the abdominal midline (linea alba), in between the knees. The dosage is weight-dependent. See instructor for the recipe and the dosage.

Copyright © 2015, 2011, 2008 Pearson Education, Inc.

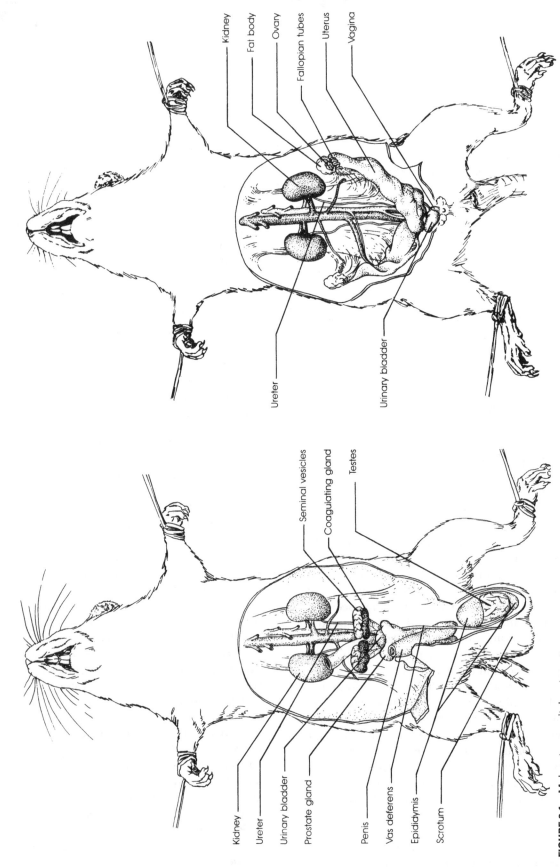

FIGURE 9.2   Female rat urogenital system.

FIGURE 9.1   Male rat urogenital system.

Copyright © 2015, 2011, 2008 Pearson Education, Inc.

⚠ **CAUTION!**

Watch the depth of respiration continuously so it does not become too shallow. If your animal starts to awaken, cover the nose with the isoflurane cone until it is, once again, sedated. Do not apply the cone continuously.

6. After the operation, place the animals in a box containing clean towels until they are fully conscious. A lamp placed over the box will help them regain the body heat they lost during the operation. Periodic massage also helps in their recovery. Later, they will be returned to their clean home cage.

7. While still anesthetized, number the rats for positive identification later. Tail markings or ear tags are the best means of identification.

CLINICAL APPLICATION

Anabolic Steroid Hormones and Athletic Performance

Anabolic steroids or anabolic-androgenic steroids (AAS) are used to enhance athletic performance and are universally banned by the governing bodies of all types of sports. The anabolic effects of these hormones include an increase in protein synthesis, bone marrow, and the resultant production of red blood cells, remodeling, and growth in the size of skeletal muscles. Androgenic effects include pubertal growth, sebaceous gland oil production, growth of the clitoris in females and the penis in male children (not adults), increased size of the vocal cords, and an increase in androgen-sensitive hair. Adverse effects of anabolic steroid use include increased acne, altered glucose metabolism, increased risk of coronary artery disease and cardiovascular disease, liver damage, an increase in LDL cholesterol and a decrease in HDL cholesterol, and mood disorders. Sex-specific side effects include gynecomastia, testicular hypertrophy, and reduced fertility in males. In females, the side effects can include increased body hair, a deeper voice, an enlarged clitoris, and irregular and reduced menstrual cycles.

► **ACTIVITY 9.1**

Testicular and Gonadotropic Hormones

Materials (per group)

☐ 4 male rats, 75–100 g in weight

☐ 1 each forceps and scissors (sterile)

☐ Betadine antiseptic

☐ 95% ethyl alcohol

☐ 1 precision balance (0.1 g readability)

☐ Isoflurane

☐ Ketamine-xylazine anesthetic cocktail

☐ 1 1-ml insulin syringe

☐ 1 dessicator

☐ Wound clips

☐ CO_2 tank and regulator or isoflurane (to euthanize the rats)

☐ Thick string

Male rats weighing 75 to 100 g are divided into the following categories:

☐ Normal rat (control)

☐ Rat castrated 14 days prior to the last lab

☐ Rat castrated 14 days prior to the last lab and given subcutaneous injection of 0.1 mg of testosterone daily for the last 10 days

☐ Normal rat given a subcutaneous injection of chorionic gonadotropin (20 units) daily for the last 10 days

☐ Rat castrated 14 days prior to the last lab and given a subcutaneous injection of chorionic gonadotropin (20 units) daily for the last 10 days

Record initial and final body weights for all rats in the table in the Laboratory Report. On the final day, you will sacrifice the rats using an overdose of isoflurane or CO_2 in a desiccator. You will carefully dissect out and weigh the seminal vesicles and record their weight in milligrams per 100 g of body weight. Students will then present their findings to the class and discuss the results.

Directions for Removal of Testes (Castration)

1. The intraperitoneal anesthetic cocktail is designed to keep the animal anesthetized for 30–40 min. The animal is unaware of the operation and should not move when the skin is cut. If the animal shows signs of awakening during the operation, place an isoflurane cone over its nose and mouth. Remove the cone when the animal appears to go under again. Remember to replace the dry cotton wad to prevent the escape of isoflurane into the general laboratory space.

2. Clip the hair along the ventral midline of the scrotum with scissors and swab the area first with betadine and then with alcohol. Using fine-pointed scissors, make a midline incision about 1.5 cm long through the scrotal skin.

3. Sometimes the testes retract into the abdominal cavity and, therefore, are not visible in the scrotal sac. Slight pressure on the lower abdominal area will force the testes back into the scrotum.

Copyright © 2015, 2011, 2008 Pearson Education, Inc.

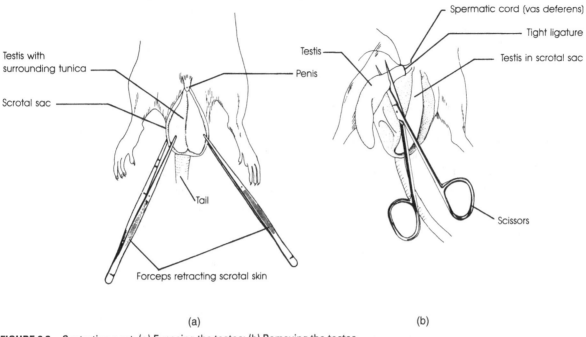

FIGURE 9.3 Castrating a rat. (a) Exposing the testes; (b) Removing the testes.

You will note that each testis is surrounded by a translucent membrane called the tunica (Figure 9.3a). Grasp the tunica with forceps and slit it with the scissors to free the testis.

4. Tie a heavy thread tightly around the vas deferens (spermatic cord) and surrounding fat tissue; then cut the spermatic cord between the knot and the testis and remove the testis (Figure 9.3b). Remove the other testis in the same manner.

5. Close the skin incision, using two or three wound clips. Place the rat in the recovery room box until it regains consciousness. It is sometimes necessary to massage the animal to facilitate recovery. ◄

► ACTIVITY 9.2

Ovarian Hormones and Estrus Cycle

Materials (per group)

- ☐ 3 female rats, 75–100 g each
- ☐ 1 each forceps and scissors (sterile)
- ☐ 1 precision balance (0.1 g readability)
- ☐ Isoflurane
- ☐ Ketamine-xylazine anesthetic cocktail
- ☐ 1 1-ml insulin syringe
- ☐ 1 dessicator

- ☐ Silk thread sutures
- ☐ Wound clips
- ☐ Ball cotton
- ☐ Toothpicks
- ☐ Slides and coverslips
- ☐ Giemsa stain
- ☐ Betadine antiseptic
- ☐ 95% ethyl alcohol
- ☐ CO_2 tank and regulator or isoflurane (to euthanize the rats)

Female rats weighing 75 to 100 g are divided into the following categories:

- ☐ Normal (control)
- ☐ Rats ovariectomized 14 days prior to the last lab
- ☐ Rats ovariectomized 14 days prior to the last lab and given a subcutaneous injection of 0.1 mg of estrogens daily for the last 10 days

Record initial and final body weights for all rats in the table in the Laboratory Report. On the final day, you will sacrifice the rats, using an overdose of ether in a desiccator. You will dissect out the entire uterus and record the weight in milligrams per 100 g of body weight.

Starting three days after surgery, you will take daily vaginal smears from each rat to ascertain the condition of the estrus cycle in each animal. After the last lab, students will present their findings to the class and discuss the results.

Copyright © 2015, 2011, 2008 Pearson Education, Inc.

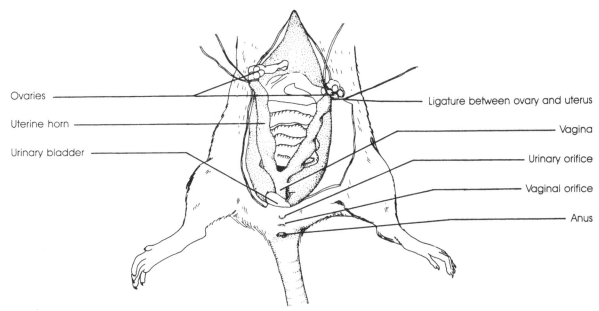

FIGURE 9.4   Preparing the rat for ovariectomy.

1. Directions for Removal of Ovaries (Ovariectomy)

a. Anesthetize the rat and make a ventral midline incision into the abdominal cavity. If you make this incision through the linea alba (connective tissue in abdominal midline), you will obtain a bloodless field. Be careful not to cut the diaphragm.

b. Push aside the intestines and locate the two horns of the bicornate uterus. Follow the uterus forward until the ovary comes into view.

c. Place a tight ligature around the uterus just below the ovary (Figure 9.4), then cut through the uterus on the ovarian side of the tie and remove the ovary. Repeat this procedure for the other ovary.

d. Close the muscle incision with silk thread sutures and the skin incision with wound clips.

e. Allow three days for recovery from the operation and then take vaginal smears daily for detection of the estrus cycle.

2. Determination of the Estrus Cycle by Using Vaginal Smears

The term **estrus** means frenzy. This is the time in the rat's reproductive cycle when she is receptive to sexual copulation and ovulation occurs. We sometimes say that the animal is in heat in this period. The estrus cycle in the rat lasts about five days and is repeated throughout the year (polyestrus). The cycle is typically divided into the following four stages, which can be identified by examining changes in the types of cells lining the vagina.

Proestrus This is a period of increasing levels of **follicle-stimulating hormone** (FSH) and **luteinizing hormone** (LH), which stimulate follicle growth and secretion of estrogens. Vaginal smears contain mainly nucleated epithelial cells. Proestrus lasts 8 to 12 hr.

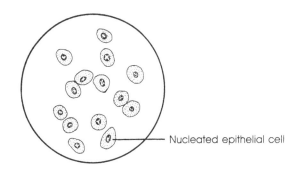

Estrus This is the period of heat and copulation. High levels of estrogens stimulate mitosis of cells in the uterus and vagina. Vaginal smears contain many cornified cells and few leukocytes or nucleated epithelial cells. Estrus lasts 9 to 15 hr.

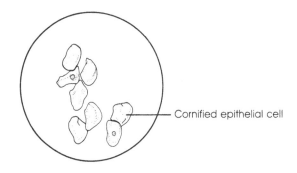

Copyright © 2015, 2011, 2008 Pearson Education, Inc.

Metestrus LH and **luteotropic hormone** (LTH) promote the formation of the corpus luteum in this period, which lasts for 10 to 14 hr. The secretion of both progesterone and estrogens increases. Vaginal smears contain many leukocytes and some cornified cells.

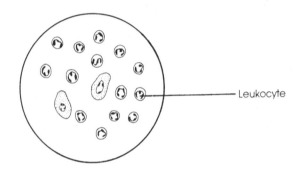

Leukocyte

Diestrus This is the longest stage, lasting 60 to 70 hr. The corpus luteum regresses, and the uterus is small and poorly vascularized. Levels of gonadotropic and sexual hormones are at low levels. Vaginal smears contain mainly leukocytes.

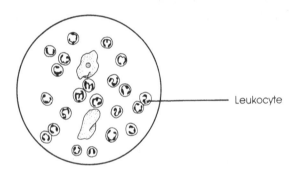

Leukocyte

Carry out the following steps to complete the experiment:

a. Grasp the rat behind its jaw, holding it on its back in the palm of your hand. Wrap a tuft of cotton tightly around a toothpick, moisten the swab with saline, and *gently* insert and rotate the swab within the vagina. Then press the swab in a drop of saline on a microscope slide and smear the cells evenly over the slide.

b. To stain the cells, place the air-dried slide in methyl alcohol for 5 sec, remove it, and air dry it. Then place the slide in Giemsa stain for 15 min, remove it, rinse it with distilled water, and air dry it before examining it under the microscope.

c. Determine the stage of the estrus cycle for each rat for each of the 10 days prior to the last lab. **Enter your results in the table in the Laboratory Report.** ◀

Pregnancy Tests

Tests for human pregnancy are a good demonstration of some endocrinologic principles and are relatively easy to do. There are several types of pregnancy tests, all of which are based on the detection of **chorionic gonadotropin**, which is produced in the female during early pregnancy.

After the fertilized ovum is implanted in the uterine wall, trophoblastic cells around the ovum begin producing chorionic gonadotropic hormones. (Trophoblastic cells differentiate into a part of the placenta called the chorion, hence the name *chorionic* gonadotropin.) These hormones closely resemble the anterior pituitary gonadotropic hormones in activity, having both luteinizing and luteotrophic properties. Hence, they promote the growth and maintenance of the corpus luteum to ensure adequate levels of estrogen and progesterone in early pregnancy. The production of **human chorionic gonadotropin** (hCG) increases sharply as the placenta develops and reaches a peak level approximately eight weeks after the first day of the last menstrual period. In early pregnancy, then, the hCG level becomes so elevated that it spills over into the urine and, thus, the detection of hCG in the urine becomes the basis for the pregnancy tests.

● STOP AND THINK

Human chorionic gonadotropin is released by a zygote and mimics the action of LH, causing the corpus luteum to persist and maintain the high levels of estrogen and progesterone necessary to prevent the uterine lining from shedding. Although hCG levels can vary significantly between women, extremely high or low levels of hCG can be cause for concern. Very low levels of hCG can indicate a possible miscarriage, an ectopic pregnancy, or a miscalculation of the date of conception. Very high levels might indicate a molar pregnancy (where only placental tissue but no fetus is developing), multiple fetuses, or a miscalculation of the date of conception. The hCG hormone, due to its LH-like action, also increases testosterone production in males and masks anabolic-androgenic steroids (AAS) use.

There are two classes of pregnancy tests: **biological** and **immunological**. The biological tests depend on the effect of the chorionic gonadotropin from the woman being tested on the reproductive organs of test animals. The Ascheim–Zondek test is based on the fact that hCG stimulates immature ovaries to cause ovulation and to secrete hormones that stimulate uterine growth.

Gonadotropin can also be detected by immunologic reactions. Chorionic gonadotropin is a protein and behaves as an antigen when injected into an

Copyright © 2015, 2011, 2008 Pearson Education, Inc.

animal. The animal becomes immune to the chorionic gonadotropin by producing antichorionic gonadotropin **antibodies**. The blood serum of the injected animal will contain these antibodies, which are, therefore, called **antiserum**.

▶ ACTIVITY 9.3
Pregnancy Tests

Consistent, positive tests can be obtained by preparing synthetic pregnant urine made of 0.9% NaCl containing 100 IU/ml of chorionic gonadotropin, which can be considered pregnant urine. Use 0.9% saline without hCG to represent nonpregnant urine.

1. Immunological Tests for Pregnancy

Materials

☐ 1 commercially available home pregnancy test strip

☐ 10 ml pregnant urine

☐ 10 ml nonpregnant urine

Home pregnancy tests use hCG-specific antibodies to detect the presence of hCG in the urine of a pregnant woman. Although the sensitivities of home pregnancy tests vary, most are sensitive enough to detect pregnancy within 10 to 12 days after conception. The most commonly available commercial tests use a test strip embedded with anti-hCG antibodies. The combination of anti-hCG with the hCG in the urine of a pregnant woman triggers a reaction that causes a color change, indicating a positive response. The absence of a color change indicates the absence of hCG, a negative response.

a. Obtain samples of pregnant and nonpregnant urine in separate sample cups.
b. Label each cup appropriately.
c. Follow the manufacturer's instructions to test the urine by using separate sample sticks for each cup.
d. Observe the difference in the response.

2. Ascheim–Zondek Pregnancy Test

Materials

☐ 1 immature female rat or mouse

☐ 1 1-ml insulin syringe and needle

☐ 10 ml pregnant urine

☐ 10 ml nonpregnant urine

☐ 1 dissecting microscope or magnifying lens

☐ 1 each forceps and scissors

The Ascheim–Zondek test is a biological test for pregnancy that was widely used before the development of the immunological test. It is based on the effect of chorionic gonadotropin on the ovaries and uterus of mice, rats, and rabbits.

a. Select immature female rats or mice, 15 to 25 days old (weighing 40–45 g). You will give half of the animals a subcutaneous injection of pregnant urine twice daily (morning and evening) for three days prior to lab, and you will give the other half injections of normal (nonpregnant) urine on the same schedule. For rats, inject 1 ml of urine each time; for mice, inject 0.5 ml of urine.
b. Four to five days after the first injections, sacrifice and examine the rats.
c. First, examine the vaginal orifice (opening) for any differences in size or patency between the experimental groups. Explain any difference.
d. Next, open the abdominal cavity and examine the uterus and ovaries of the animals. Compare the uteri of the pregnant urine–injected and control animals for differences in size and vascularity. Examine the ovaries carefully, using a hand lens or a dissecting microscope. The presence of hemorrhage spots (corpora hemorrhagica) or yellow bodies (corpora lutea) indicates that eggs have been released—their presence is considered a positive test for pregnancy. ◀

Copyright © 2015, 2011, 2008 Pearson Education, Inc.

Digestion

CHAPTER 10 INCLUDES:

PhysioEx™ 9.I For more exercises on Digestion, visit PhysioEx™ (www.physioex.com) and choose Exercise 8: Chemical and Physical Processes of Digestion.

OBJECTIVES

After completing this exercise, you should be able to

1. Demonstrate the role of enzymes in digestion.
2. Identify the pH and temperature optima for different locations in the digestive tract.
3. Explain the role of bile salts in the digestion of fats.

Living organisms run on energy, and it is the job of the digestive system to reduce the foods we eat to small molecules the cells can use to produce adenosine triphosphate (ATP). This degradation process is catalyzed by hydrolytic enzymes, which split large molecules into smaller units by combining with water. The result of digestion is the reduction of carbohydrates to monosaccharides, proteins to amino acids, and fats to fatty acids and glycerol. Hydrolytic reactions are made more efficient by the division of the digestive tract into compartments where specific enzymes can operate at their optimum pH. Movement between these compartments requires coordinated muscle contractions such as those seen in peristalsis and segmentation. Secretion of enzymes within these compartments at the proper time is controlled by neural reflexes and hormones such as gastrin, secretin, cholecystokinin, and gastric inhibitory peptide. The movement of digested materials across the mucosa of the digestive tract into the bloodstream occurs through absorption. A summary of digestive function is presented in Figure 10.1.

In this exercise, you examine the action of some of the key digestive enzymes and the factors that alter their activity. As you work, use your text and lecture notes to become better acquainted with the enzymes and hormones operating along the digestive tract. Because it is important to control the conditions under which these experiments are conducted, avoid contamination by making sure that you use separate droppers and pipettes for each of the solutions.

Salivary Digestion of Carbohydrates

Digestion of carbohydrates begins in the mouth, where the salivary glands (parotid, sublingual, and submandibular) secrete an **amylase** called **ptyalin** that begins the hydrolysis of complex polysaccharides:

$$\text{Plant starches or animal glycogen} \xrightarrow{\text{Amylase}} \text{Disaccharides (maltose, dextrin)}$$

Ptyalin has an optimum pH of approximately 6.8, which is roughly the pH found in the mouth. In the following activity, you examine ptyalin digestion of starch, using **Benedict's test** to measure maltose formation and **Lugol's** (iodine) **solution** to test for starch.

▶ ACTIVITY 10.1
Salivary Digestion of Carbohydrates

Materials (per group)

- ☐ 4 20-ml test tubes
- ☐ 20 ml .5% starch paste
- ☐ 20 ml distilled water
- ☐ 20 ml salivary amylase (400 units per 100 ml)
- ☐ 10 ml 1% acetic acid
- ☐ 10 ml concentrated HCl
- ☐ 10 ml Lugol's solution
- ☐ 10 ml Benedict's solution
- ☐ Ice bath
- ☐ Water bath at 37°C
- ☐ Test tube stand

Copyright © 2015, 2011, 2008 Pearson Education, Inc.

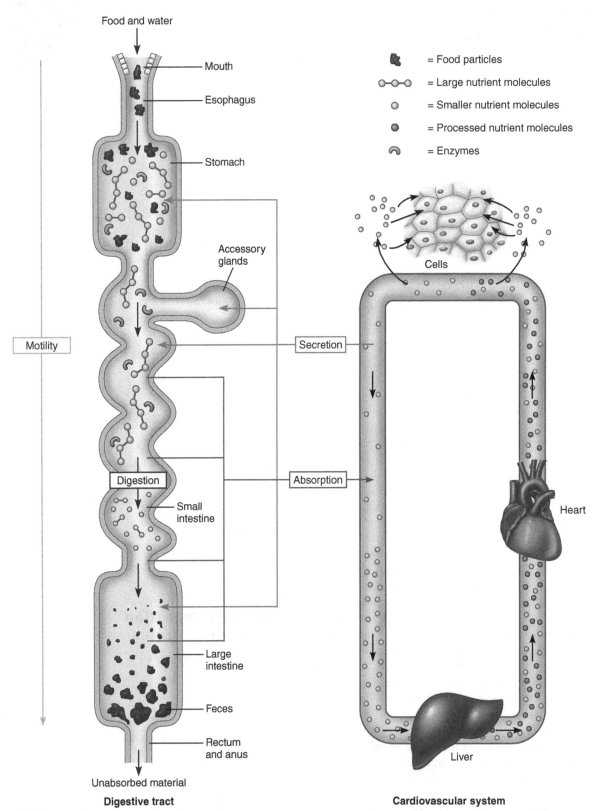

FIGURE 10.1   An overview of digestive function.

Source: Stanfield, Cindy L.; Germann, William J., *Principles of Human Physiology*, 3rd Ed., © 2008, pg. 571. Reprinted and electronically reproduced by permission of Pearson Education, Inc., Upper Saddle River, New Jersey.

Copyright © 2015, 2011, 2008 Pearson Education, Inc.

1. Prepare a solution of salivary amylase in water (400 units per 100 ml). This will substitute for human saliva.
2. Place a small amount of the solution in a spot plate and add a few drops of 1% acetic acid. A precipitate indicates that mucin (a glycoprotein) is present.
3. Prepare and label four test tubes as follows (use a 0.5% starch paste):

Tube 1	**Tube 2**
3 ml starch	3 ml starch
+	+
3 ml water	3 ml saliva
↓in	↓in
37°C water bath	37°C water bath

Tube 3	**Tube 4**
3 ml starch (cooled)	3 ml starch
+	+
3 ml saliva (cooled)	3 ml saliva
↓in	+
ice bath	5 drops conc. HCl
	↓in
	37°C water bath

4. After the tubes have incubated for 1 hr, pour half of each tube's contents into a new test tube. Test one set of tubes for starch, using Lugol's solution, and the other set for maltose, using Benedict's solution. **Enter your observations and answer the questions in the Laboratory Report.**
   a. *Starch test* Add three drops of Lugol's solution to each tube. A dark purple color indicates the presence of starch. Shades of reddish brown indicate lesser amounts of starch. Rate the amount of starch as (+++), (++), (+), or (−). A dark purple result would indicate that no starch digestion has taken place, whereas a lighter reddish brown coloration would indicate that most of the starch has been digested.
   b. *Maltose test* Add 4 ml of Benedict's solution to each tube and place in a boiling water bath for 2 min. Remove the tubes by using a clamp and compare the concentration of maltose, using the following scale: (+++), red; (++), orange-yellow; (+), green; (−), blue. A red result would indicate that most of the starch has been converted to maltose, whereas a blue coloration would indicate that maltose has not been formed. ◄

Gastric Digestion of Protein

Protein digestion begins in the stomach, where the pepsin enzyme splits proteins into shorter polypeptide chains containing amino acids. Secretion and activation of pepsin occur as follows:

$$\text{Chief cells in gastric pits} \rightarrow \text{Pepsinogen} \xrightarrow{+\text{HCL}} \text{Pepsin}$$

The strong hydrochloric acid secreted by the parietal cells has two functions in gastric digestion: It activates pepsin, and it produces a stomach pH of about 2, which is optimal for pepsin activity. Activity 10.2 illustrates some of the features of protein digestion by pepsin.

EXPLAIN THIS!
In addition to the two functions of HCl listed in the paragraph above, what other non-digestive function does HCl play?

CLINICAL APPLICATION

Peptic Ulcer Disease (PUD)

Peptic ulcer disease (PUD) is probably the most common disease associated with the digestive system. Inflammation of the gastric and duodenal mucosa can result in symptoms that include epigastric pain, nausea, pernicious anemia, and life-threatening hemorrhage. Although the gastric and duodenal linings are normally protected from ulcerations by the mucus and prostaglandins found in these areas, infections of *Helicobacter pylori* or overuse of NSAIDs (nonsteroidal anti-inflammatory drugs) such as aspirin and ibuprofen overcome these natural defenses and could lead to ulceration. Depending on the cause, antibiotics to combat *H. pylori* infection or using alternate NSAIDs will usually treat PUD successfully.

▶ ACTIVITY 10.2
Gastric Digestion of Protein
Materials (per group)

☐ 4 20-ml test tubes	☐ 10 ml 1% acetic acid
☐ 1 scalpel	☐ 10 ml 0.5% NaOH
☐ 20 ml distilled water	☐ 10 ml 0.5% HCl
☐ 20 ml 5% pepsin solution	☐ Water bath at 37°C
☐ 1 hard-boiled egg	☐ Test tube stand

1. Place thin slices of cooked egg white in four test tubes. It is important to make these slices the same size (about 0.5 cm^2) and as thin as possible.

Copyright © 2015, 2011, 2008 Pearson Education, Inc.

2. Add the following solutions to the test tubes and determine the pH of each tube:

Tube 1	Tube 2
5 ml pepsin	5 ml pepsin
(5% soln)	(5% soln)
+	+
5 ml HCl	5 ml water
(0.5%)	

Tube 3	Tube 4
5 ml HCl	5 ml pepsin
(0.5%)	(5% soln)
+	+
5 ml water	5 ml NaOH
	(0.5%)

3. Allow the tubes to incubate in a 37°C water bath for 1 hr. Test the final pH of the solutions and estimate the amount of protein digestion visually by subjectively comparing the size of the egg slice in each situation. Use a scale of (+++) for the smallest remnant and the most digestive activity and (++), (+), and (−) for the largest remnant and the least digestion to compare the four tubes. **Enter your observations and answer the questions in the Laboratory Report.** ◄

Digestion of Fat with Pancreatic Lipase and Bile Salts

Pancreatic lipase has a major role in fat digestion, but by itself, lipase is ineffective because it is a water-soluble enzyme trying to act on large lipid droplets, which are water insoluble. Bile salts help overcome this problem by acting as emulsifying agents that break the fat into smaller droplets so that lipase has a larger surface area for the hydrolysis of fats. Bile salts are amphipathic molecules, having one region that is hydrophilic and attracted to the surrounding water. The other region is hydrophobic and associates with the surface of the lipids. A mixture of bile salts with fat and water will create a stable emulsion of small fat droplets in water, thereby facilitating the action of lipase. Pancreatic colipase is also needed to displace the bile salts covering the fat droplets to allow lipase activity within the bile salt coating. When lipase is added to any fat, its digestive activity can be recognized by the production of acid (indicated by a change in pH). Butyric acid is produced when lipase digests dairy fat, and it is easily recognized by a characteristic rancid odor.

The pancreas also aids digestion by secreting sodium bicarbonate. This compound produces a pH of approximately 7.8 in the small intestine, which is optimal for the action of the pancreatic enzymes. In the following activity, you examine some aspects of the action of pancreatic lipase and bile salts on lipids.

▶ ACTIVITY 10.3

Digestion of Fat with Pancreatic Lipase and Bile Salts

Materials (per group)

- ☐ 4 20-ml test tubes
- ☐ 20 ml distilled water
- ☐ 20 ml 1% pancreatin solution
- ☐ 20 ml fresh dairy cream
- ☐ 5 gm bile salts
- ☐ Litmus powder
- ☐ Water bath at 37°C
- ☐ Test tube stand

Emulsification

1. In each of two test tubes (labeled A and B), place 3 ml of distilled water and 3 ml of vegetable oil. To tube B, add a small pinch of bile salts. Shake each tube for 30 sec and observe it for several minutes. **Enter your observations and answer the questions in the Laboratory Report.**

Digestion

2. Add litmus powder in small amounts to dairy cream until a blue color is produced. Preincubate the litmus cream and a 1% pancreatin solution at 37°C for 5 min. Prepare a series of test tubes as follows:

Tube 1	Tube 2
3 ml cream	3 ml cream
+	+
3 ml pancreatin	3 ml water

Tube 3	Tube 4
3 ml cream	3 ml cream
+	+
3 ml pancreatin	3 ml water
+	+
pinch bile salts	pinch bile salts

3. Incubate all tubes in a 37°C water bath for 1 hr or until a color change occurs in one tube. Blue litmus will turn pink in an acid environment. Use pH paper to test the pH and note the odor of each tube. **Enter your observations and answer the questions in the Laboratory Report.** ◄

Copyright © 2015, 2011, 2008 Pearson Education, Inc.

INQUIRY-BASED ACTIVITY

Appendix C describes the format of a typical Laboratory Report. It is mandatory that you read the Appendix before you start planning your experiment.

Each of the experiments described above can be modified to test the effects of temperature or pH on the digestive activity of the enzyme involved.

For activities 10.1–10.3, use your research to determine which experimental setup would best allow you to test the effects of varying temperatures, such as an ice bath, 20°C water bath, 37°C, and a 50°C water bath on any of the enzymes used. Be sure to describe your rationale in your Laboratory Report and also how you would estimate digestive activity.

Similarly, one can test the effects of pH on the activity of the enzymes concerned in activities 10.1–10.3. You can use varying amounts of HCl to create environments of different acid levels. Varying amounts of NaOH can be used to create different alkaline environments.

An appropriately modified table, modeled after that shown in the Laboratory Report, can be used to present your data.

Complete this checklist to make sure that you have covered all bases before you start your experiment:

What is the question you seek to answer?

Frame it in the form of a hypothesis.

What is the independent variable or treatment?

How will the independent variable or treatment be measured?

What is the control treatment?

How will you replicate your experiment?

How will you ensure that your subjects are similar enough to not introduce some other independent variability? Are there any standardized variables?

What is/are the dependent variable(s)?

How will the dependent variable(s) be measured?

What are your predictions?

Construct a table to record your observations easily.

How will you present the data collected graphically?

How will you analyze the data collected?

How will you know if the differences between the treated and the untreated samples are statistically significant?

Your laboratory instructor will describe how your Laboratory Report will be written.

Copyright © 2015, 2011, 2008 Pearson Education, Inc.

LABORATORY REPORT 10

Digestion

Name _Caroline Leibel_

Date _____ Section _17_

Score/Grade _____

Salivary Digestion of Carbohydrates

1. pH of saliva _____ Mucin present? _____
2. What is the function of mucin in the mouth? _____
3. Indicate the relative amounts of starch and maltose after incubation:

Tube	Starch	Maltose	Explanation
1: Water	Yes	No	No digestion
2: Saliva	No	Yes	Digestion is best here b/c it has starch, enzyme & ideal temp
3: Cooled Saliva	No	Yes	Digestion occurs but not ideal temp
4: Saliva, HCl	Yes	No	Small amount of digestion b/c HCl makes pH not ideal

4. What *in vivo* (in the body) situation is simulated by the conditions in tube 4? _____
5. Does ptyalin hydrolysis of carbohydrate continue in the stomach?
   Explain. _____
6. Where else is amylase secreted in the digestive system? _____
   Pancreatic juice and the small intestine (microvilli of epithelial cells)

Gastric Digestion of Protein

1. Record your results on egg white digestion.

Tube	Initial pH	Final pH	Estimated Digestion	Explanation
1: Pepsin, HCl	—	—	1	HCl creates optimal pH, contains pepsin
2: Pepsin, water	—	—	2	pepsin will sort of work, but water does not create optimal pH
3: HCl, water	—	—	4	No digestion will take place b/c no pepsin
4: Pepsin, NaOH	—	—	3	Some digestion will occur, but not optimal pH

2. What *in vivo* situation is simulated by tube 4? _Chyme entering_
3. Which other enzymes have major proteolytic activities in the digestive tract? _____
4. A person with achlorhydria has defective secretion of HCl by the parietal cells. What is the physiological effect of achlorhydria in the body? _____
   _____
5. What is the function of the mucous cells in the gastric pits? _____

Digestion of Fat with Pancreatic Lipase and Bile Salts

Emulsification

1. Which tube (A or B) has the smaller and more dispersed fat droplets? _____
2. What are bile salts? _____
   What are bile pigments? _____

Copyright © 2015, 2011, 2008 Pearson Education, Inc.

3. Where is bile secreted? _____

4. On a separate sheet of paper, illustrate the mechanism of bile salts in the emulsification process.

Digestion

5. Record the final color, pH, and odor of each tube involved in the digestion of cream.

Tube	Color	pH	Odor	Explanation
1: Pancreatin	Pink	—	Minor cheese	No bile salt
2: Water	Purple	—	Nothing	Don't have pancreatin or bile salts
3: Pancreatin, bile salts	Pink	—	Major Cheese	Bile salts accelerate digestion w pancreatin
4: Water, bile salts	Purple	—	Nothing	No Pancreatin

6. What produces the acid pH, indicating that fat digestion has occurred? *Release of fatty acids*

7. What produces the rancid odor with fat digestion? _____

8. Which enzymes are present in the pancreatin solution? _____

9. Which enzymes are present in the microvilli brush border of the small intestine? _____

10. Use your lecture textbook to complete this table on gastrointestinal hormones.

Hormone	Site of Origin	Release Stimulus	Function
Gastrin			
Secretin			
Cholecystokinin			
Gastric inhibitory peptide			

11. How does the intestinal absorption of lipids differ from the absorption of glucose and amino acids?

_____

_____

APPLY WHAT YOU KNOW

1. Why aren't the acinar cells of the pancreas digested by the proteolytic enzymes they secrete?

Because they are inactive when secreted, but then activated when they enter the intestinal lumen

2. Heartburn is commonly treated with buffers such as sodium bicarbonate and aluminum hydroxide, which help neutralize the effects of HCl in the stomach. Newer therapies such as H_2-receptor antagonists or proton pump inhibitors also help alleviate the symptoms of heartburn. Use your knowledge of stomach physiology to explain how each of these medications works.

_____

_____

_____

Copyright © 2015, 2011, 2008 Pearson Education, Inc.

Smooth Muscle Motility

⚠ CAUTION!

Because these experiments involve the use of living vertebrates, it is important that you obtain the appropriate permissions from your Institutional Animal Care and Use Committee (IACUC). While the instructions given are suitable to obtain the desired result, your IACUC may require a modification of procedures to meet your institution's needs.

OBJECTIVES

After completing this exercise, you should be able to
1. Demonstrate the properties of smooth muscle contraction.
2. Demonstrate the effect of hormones on smooth muscle contraction.

Responses of Intestinal and Uterine Segments

Smooth muscle is a non-striated involuntary type of muscle found in the stomach, intestines, bladder, and uterus and in the walls of most arteries and veins. Its activity is controlled by the autonomic nervous system, but all smooth muscle does not respond in the same manner to the same stimulus. For example, sympathetic stimulation produces a decrease in the rhythmic contractility of the gut smooth muscle but causes the circular smooth muscle of the arteries to contract and constrict the vessels. Parasympathetic stimulation evokes an increase in intestinal contractility but has little effect on arterial smooth muscle. Another unique property of smooth muscle is its ability to maintain approximately the same amount of tone and rhythmic contractility at different degrees of stretch. Because of this property, the intestines, uterus, and bladder can undergo "receptive relaxation." For example, the uterus can be stretched greatly to receive a growing fetus without increasing its muscular tension on the fetus or its contractility. In the following activity, you will use an isolated segment of the intestine and uterus to demonstrate some of these special properties of smooth muscle.

⬤ STOP AND THINK

In most cases the neurotransmitters released by sympathetic and parasympathetic neurons produce antagonistic responses in smooth muscle. What is the physiological basis for these differences?

▶ ACTIVITY 11.1

Smooth Muscle Response

Materials (per group)

☐ 1 150–200 g female rat

1. Prepare a muscle warmer setup complete with aeration as shown in Figure 11.1. During the activity, place the muscle warmer in a 37°C water bath to keep the smooth muscle at its normal physiological temperature. Reserve flasks containing Locke's solution should also be placed in the water bath so that they will be at the proper temperature when needed.
2. Give no food to the rats used in this experiment for 24 hr prior to the lab period to remove most of the fecal matter from the intestinal lumen.
3. Kill a female rat (150–200 g) using an overdose of CO_2 in a desiccator. (Many anesthetic agents such as ether or Nembutal depress intestinal motility, whereas CO_2 has little effect.)
4. Open the abdominal cavity and isolate a 10-cm segment of the jejunum. The jejunum begins about 8 cm below the gastroduodenal junction (pyloric sphincter). Isolation and removal of the segment must be done gently; avoid excessive stretching or drying of the tissue. Trim off the mesentery and place the segment in a dish containing aerated Locke's solution at 37°C. Flush out the gut contents with Locke's solution using gentle pressure from a syringe into the lumen of the jejunum segment.
5. Also at this time, remove the entire uterus (both horns) and place it in an aerated Locke's solution (37°C) for use later in the activity.

Copyright © 2015, 2011, 2008 Pearson Education, Inc.

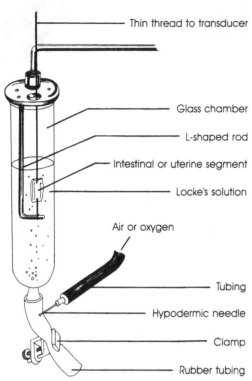

- Thin thread to transducer
- Glass chamber
- L-shaped rod
- Intestinal or uterine segment
- Locke's solution
- Air or oxygen
- Tubing
- Hypodermic needle
- Clamp
- Rubber tubing

FIGURE 11.1 Smooth muscle warmer setup.

6. Cut the jejunum segment into smaller segments about 2 cm long. Tie one end of a segment to the L-shaped rod of the muscle warmer and the other end to the transducer (use a thin thread). Use the procedures described in Chapter 15 to measure muscle responses.

7. Fill the muscle warmer with Locke's solution so that the segment is completely immersed. Turn on the air or oxygen so that it gently bubbles through the solution. Adjust the transducer sensitivity so that normal contractions about 2 cm high are recorded.

8. Record the contractions of the jejunum segment under each of the experimental conditions listed in the table in the Laboratory Report. You will need to drain the muscle warmer and refill with fresh Locke's solution periodically as indicated.

9. After you are finished with the jejunum segment, mount one horn of the uterus in the muscle warmer and record its contractility under the conditions given in the table in the Laboratory Report.

10. Save your myographs for insertion in the Laboratory Report.

Follow the sequence of activities described in the table in the Laboratory Report to complete this experiment. Be sure to change solutions when directed to avoid compromising the results. ◄

Copyright © 2015, 2011, 2008 Pearson Education, Inc.

Smooth Muscle Motility

Name _____

Date _____ Section _____

Score/Grade _____

Responses of Intestinal and Uterine Segments

1. Observations on contractility of smooth muscle

Experimental Conditions	Frequency	Contractile Strength	Tone	Explanation
Jejunum Segment				
Normal activity				
Increased stretch (raise transducer or add counterweights)				
Epinephrine: add 1 or 2 drops of 1:10,000				

Drain muscle warmer and add new Locke's solution

Acetylcholine: add 4 or 5 drops of 1:10,000				
Pilocarpine: add 2 drops of 1:1,000				
Atropine sulfate: add 2 drops of 1:4,000				
Pilocarpine: 5 min after atropine dose				

Drain muscle warmer and add new Locke's solution

$BaCl_2$: add a few drops of 0.6% soln				
HCl: add several drops of 2% soln				
NaOH: add several drops of 2% soln				
Isosmotic glucose: drain warmer and fill with 5.4% glucose soln				

Copyright © 2015, 2011, 2008 Pearson Education, Inc.

Experimental Conditions	Frequency	Contractile Strength	Tone	Explanation
Uterine Segment				
Normal activity				
Acetylcholine: add 4 or 5 drops of 1:10,000				
Epinephrine: add 1 or 2 drops of 1:10,000				

2. Myographs of smooth muscle motility

   Place the records obtained in the following spaces.

   Intestine—normal   Intestine—epinephrine

   Intestine—acetylcholine   Intestine—pilocarpine

   Intestine—atropine + pilocarpine   Uterus—normal

   Uterus—acetylcholine   Uterus—epinephrine

APPLY WHAT YOU KNOW

1. Based on the response of smooth muscle to pilocarpine, do you consider it to be an acetylcholine agonist or antagonist?

   _____

   _____

   _____

2. Based on the response of smooth muscle to atropine sulfate, do you consider it to be an acetylcholine agonist or antagonist?

   _____

   _____

   _____

Copyright © 2015, 2011, 2008 Pearson Education, Inc.

Regulation of Blood Glucose

12

CHAPTER 12 INCLUDES:

PhysioEx™ 9.1   For more exercises on Regulation of Blood Glucose, visit PhysioEx™
(www.physioex.com) and choose Exercise 4: Endocrine System Physiology.

⚠ CAUTION!

Parts of this lab might involve working with human blood. You should handle only your own blood. Dispose of all supplies (cotton, gauze, lancets, and so on) that come in contact with blood in properly marked containers. ALL BODY FLUIDS AND SUPPLIES MUST BE TREATED AS POTENTIALLY INFECTIOUS. (See "Safety in the Laboratory" inside the front cover of this text.)

OBJECTIVES

After completing this exercise, you should be able to

1. Understand the terms *absorptive state* and *postabsorptive states* in reference to metabolic processes.
2. Determine feedback mechanisms involved in regulating blood glucose levels.
3. Use a glucometer to test blood glucose levels and examine changes in blood glucose levels after consuming a known quantity of glucose.

For about 3 to 4 hours after a meal, our bodies enter into an **absorptive state** during which food that has been consumed is processed by the digestive tract and absorbed into our tissues. Because energy input usually exceeds energy output at this time, we use metabolic processes to store this energy in our cells in the form of fats or glycogen. The **postabsorptive state** occurs in between meals, when our energy intake is lower than our use of energy. Complex molecules such as fats and glycogen stored in adipocytes and muscle are broken down and released into the blood to meet the energetic demands of cellular metabolic activities. A summary of two hormones involved in these activities, insulin and glucagon, is presented in Table 12.1. Although our nutritional intake is composed of fats, amino acids, nucleotides, and carbohydrates, this laboratory exercise will address only how our body regulates blood glucose levels, the preferred energy source for most of our cells.

Glucose Metabolism

Insulin is a peptide hormone secreted by the β-cells of the islets of Langerhans in the pancreas. Its principal function is to assist the transport of glucose across the cellular membrane into the cell. Glucose transportation into the cell occurs when insulin causes the insertion of GLUT4 channels into the plasma membrane of primarily cardiac and skeletal muscle cells. This allows glucose movement into the cells through facilitated diffusion. The normal concentration for blood glucose is 90 mg/100 ml (dl) of blood, but it can range from 60 mg/dl to 140 mg/dl. Fasting blood glucose levels are typically 70 to 110 mg/dl. In a normal person, glucose levels rarely exceed 170 mg/dl in the absorptive state. When insulin is deficient or lacking, only a small amount of glucose can cross the cell membrane and be used in cellular metabolism. This low rate of transport results in excess accumulation of glucose in the blood, a condition called **hyperglycemia**. An excess of insulin causes a decrease in the level of blood glucose, or **hypoglycemia**, depending on the individual's dietary intake of glucose.

EXPLAIN THIS!

Why are fasting blood glucose levels a better indicator of pancreatic function than any randomly obtained blood glucose level?

Blood glucose levels are also regulated by glucagon, a peptide hormone secreted by the alpha cells of the pancreatic islets. The effect of glucagon is antagonistic to that of insulin. It promotes the processes of the postabsorptive state, facilitating glycogenolysis (breakdown of

Copyright © 2015, 2011, 2008 Pearson Education, Inc.

TABLE 12.1 Regulation of Metabolism by Insulin and Glucagon

	Insulin Increase	**Glucagon Increase**
Cellular glucose uptake	Increased	Decreased
Amino acid uptake	Increased	Decreased
Protein synthesis	Increased	Decreased
Fatty acid and triglyceride synthesis	Increased	Decreased
Glycogen synthesis	Increased	Decreased
Protein breakdown	Decreased	Increased
Lipolysis	Decreased	Increased
Glycogenolysis	Decreased	Increased
Gluconeogenesis	Decreased	Increased

glycogen) and gluconeogenesis (production of glucose from non-carbohydrate sources) while inhibiting the synthesis of glycogen and fat. Because the actions of these hormones are antagonistic, plasma insulin levels rise and plasma glucagon levels fall in the absorptive state, whereas plasma insulin levels fall and plasma glucagon levels rise during the postabsorptive state. The feedback mechanisms involved in regulating plasma glucose levels are summarized in Figure 12.1.

Diabetes: Types 1 and 2

The disease **diabetes mellitus** can be caused by a lack or reduction of insulin (type 1, or insulin-dependent, diabetes) or by cellular resistance to insulin (type 2, or non-insulin dependent, diabetes). Either of these will produce the typical high blood glucose level and excretion of glucose in the urine if the diabetes is uncontrolled.

Type 1 diabetes occurs when the β-cells of the islets are destroyed due to an autoimmune attack. In some instances, this attack is thought to be caused by a viral infection, but in other cases, the cause is unknown. When cells cannot metabolize glucose for energy due to the absence of insulin, they compensate by increasing their metabolism of fats and proteins. The increased metabolism of fats releases large quantities of **ketone bodies** (acetone), which are intermediate products of fat breakdown, into the blood. These are excreted in the urine and have the easily recognizable odor of acetone. Also, ketone bodies are acidic, and

their accumulation will cause a drop in blood pH; the diabetic becomes **acidotic**. Severe acidosis leads to coma and, eventually, death. Individuals who are afflicted with type 1 diabetes require daily doses of injected insulin to treat their condition.

Type 2 diabetes (the most common form of the disease) usually occurs in individuals who are older than 40 and overweight. Although insulin levels are normal or even greater than normal, the body cells do not respond adequately to the hormone by increasing GLUT4 receptors in the cell membrane. Consequently, blood glucose levels are higher than normal. Exercise does increase the amount of GLUT4 carriers in the plasma membrane of muscle fibers, reducing blood glucose levels. A diet that reduces carbohydrate intake can also reduce the effects of type 2 diabetes. If exercise and diet are inadequate, medications that increase insulin production by the pancreas or reduce glucose production in the liver are prescribed. In some cases, insulin is also prescribed for a person with type 2 diabetes.

In either case, urinary excretion of glucose (**glucosuria**) results when the concentration of blood glucose exceeds the threshold level for total reabsorption by the kidney. The increased osmolarity of the urine also causes abnormally large quantities of water to be excreted (**poly-uria**); this increased excretion of water can lead to dehydration, which in turn stimulates excessive water intake (**polydipsia**). Glucosuria, polyuria, and polydipsia are three major characteristics of diabetes.

Copyright © 2015, 2011, 2008 Pearson Education, Inc.

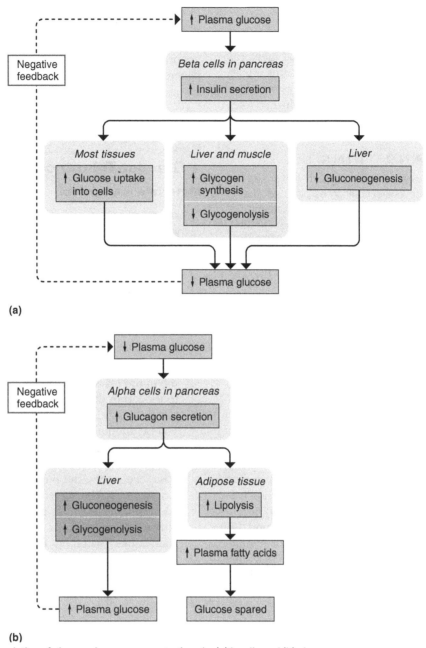

FIGURE 12.1   Regulation of plasma glucose concentrations by (a) insulin and (b) glucagon.

CLINICAL APPLICATION

Type 1 Diabetes and β-cells

Because type 1, or insulin-dependent, diabetes is a condition in which there is a lack of or reduction in the amount of insulin produced by the pancreas, current research has sought to replace the function of the absent insulin-producing β-cells. Gene therapy can use viral vectors containing the insulin sequence to infect intestinal cells that then start manufacturing insulin just like pancreatic β-cells

do. Additionally, bioengineered tissues implanted with β-cells can also be used to replace the activity of the β-cells that have been destroyed. At this time, an insulin pump can be used to deliver a predetermined dose of insulin based on "normal" expected carbohydrate intake. Because intestinal cells quickly die, bioengineered tissues can be exposed to the immune defenses of the body, and our carbohydrate intake is usually not very predictable, it is clear that much more needs to be done to fully replicate the activity of the pancreas.

Copyright © 2015, 2011, 2008 Pearson Education, Inc.

Glucose Tolerance Test

In diagnosing diabetes, several tests are used to determine as precisely as possible what metabolic error is causing the disease. Such tests are urinary glucose level, urinary ketone bodies, fasting blood glucose level, insulin sensitivity, and glucose tolerance tests.

● STOP AND THINK

When red blood cell (RBC) hemoglobin is exposed to high blood glucose levels, it becomes glycated. The hemoglobin A1c test measures the percentage of glycated red blood cells in whole blood as an indicator of blood glucose levels over the 90-day lifespan of an RBC. Hemoglobin A1c levels between 4% and 5.9% are considered normal. A1c levels are well correlated with average blood glucose levels. An A1c level of 6% (minimal for a diagnosis of diabetes) is equivalent to an average blood glucose level of 135 mg/dl, whereas an A1c level of 9% (uncontrolled diabetes) indicates an average blood glucose level of 240 mg/dl. Why would this test be a more accurate representation of a person's diabetes than would testing blood sugar levels once a day?

The glucose tolerance test assays the ability of the body (especially the pancreas) to respond to an excess ingestion of glucose. The changes in blood glucose level following glucose ingestion (1 g/kg body weight) in the normal and the diabetic person are markedly different. This difference is shown in Figure 12.2. In the normal person, the blood glucose level rises from about 90 mg/dl to around 140 mg/dl in 1 hr and then falls back to normal within 3 hr or even to below normal due to excess insulin release by the pancreas. The diabetic person, however, shows a hyperglycemic response in which the

blood glucose level rises from about 120 to 160 mg/dl to as high as 300 mg/dl and then slowly falls to the fasting diabetic level after 5 to 6 hr. The diabetic's abnormal response is caused by the inability of the pancreas to secrete additional insulin in response to elevated blood glucose levels.

▶ ACTIVITY 12.1

Glucose Tolerance Test

Materials (per group)

☐ Urine collection cups

☐ Disposable sterile lancets

☐ 500 ml 15% glucose solution in water

☐ Labstix strips

☐ Glucometer

☐ Glucometer test strips

1. Select one person from each team for this activity or ask for at least four volunteers from the entire class. These subjects should report to the lab in the fasted state (not having eaten for the past 12 to 18 hr). For our purposes, it will be adequate if they skip the meal preceding this lab.

2. Determine each subject's normal blood glucose level, using the procedure described in Activity 12.2. The subject will also obtain a specimen of his or her urine and test it for glucose using a Labstix strip.

3. Each subject will then drink 500 ml of a solution of 15% glucose (equivalent to 75 g of glucose). Popular soft drinks usually have about 50 g of sugar in a 16-oz serving.

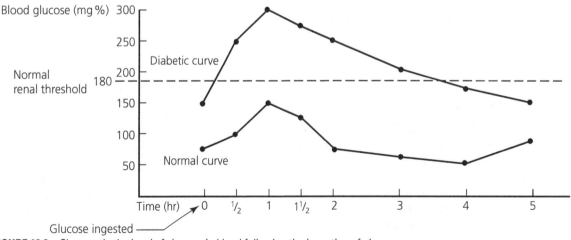

FIGURE 12.2   Changes in the level of glucose in blood following the ingestion of glucose.

Copyright © 2015, 2011, 2008 Pearson Education, Inc.

4. After ingesting the glucose or soft drink, the subject will repeat step 2 every 30 min. Testing will continue in this manner for 2 hr or until the end of the lab period.

5. **Record and graph the results of the blood glucose tests in the Laboratory Report.** Also, note the time when and if glucose appears in the urine. How do the results compare with the normal glucose tolerance test curve? **Enter your observations and graph the results of the blood glucose tests in the Laboratory Report.** ◄

► ACTIVITY 12.2

Operation of the Glucose Testing Meter

Materials

☐ Glucometer

☐ Glucometer test strips

☐ Disposable sterile lancets

An accurate measurement of blood glucose can be obtained using a blood glucose monitoring system (Figure 12.3). Many diabetics use these or other glucose meters to monitor the effect of diet, exercise, and so forth on their blood glucose so that adjustments can be made to their insulin injections or regimen of oral medication. Most are very simple to use but usually depend on the use of either an electronic coding strip or a test solution with a known glucose content to check the operation of the meter.

1. Follow the manufacturer's instructions to code your meter if necessary.

2. Remove a test strip from the carton and place it in the test slot. You will hear a beep followed by the appearance of the last test numbers or a code in the display. The meter is now ready to test your blood.

3. Wash your hands with soap and water and dry them thoroughly. Use a lancing device to prick your finger and obtain a large drop of blood. Touch the test end of the test strip with the drop of blood. The blood will be drawn in to the strip automatically. The meter will beep when sufficient blood has been obtained.

4. The test result will be displayed in mg/dl at the end of the countdown.

5. Record your result.

To begin a new test, discard the test strip and lancet in a biohazards container and use a new lancet to prick your finger. **Enter your observations in the Laboratory Report.** ◄

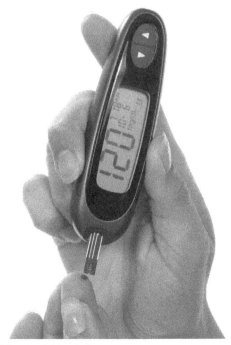

FIGURE 12.3 A blood glucose meter.

INQUIRY-BASED ACTIVITY

Appendix C describes the format of a typical Laboratory Report. It is mandatory that you read the Appendix before you start planning your experiment.

It is important that no one with diabetes participates in this exercise.

The effects of exercise on blood sugar levels has been well documented through many research studies. The insertion of GLUT4 channels into the plasma membrane of primarily cardiac and skeletal muscle cells helps in the transportation of glucose from the blood into the cells during exercise, reducing blood sugar levels. Given this, use the procedures described in Activity 12.1 to test the effects of exercise on glucose metabolism. Use the modified Harvard Step Test (described in Chapter 21) as an easily quantified measure of exercise performed.

Complete this checklist to make sure that you have covered all bases before you start your experiment:

What is the question you seek to answer?

Frame it in the form of a hypothesis.

What is the independent variable or treatment?

How will the independent variable or treatment be measured?

What is the control treatment?

How will you replicate your experiment?

Copyright © 2015, 2011, 2008 Pearson Education, Inc.

How will you ensure that your subjects are similar enough to not introduce some other independent variability? Are there any standardized variables?

What is/are the dependent variable(s)?

How will the dependent variable(s) be measured?

What are your predictions?

Construct a table to record your observations easily.

How will you present the data collected graphically?

How will you analyze the data collected?

How will you know if the differences between the treated and the untreated samples are statistically significant?

Your laboratory instructor will describe how your Laboratory Report will be written.

● STOP AND THINK

Is it important to standardize the amount of glucose consumed per individual based upon body size and gender (as described in the Inquiry-Based Activity section of Chapter 3)?

Copyright © 2015, 2011, 2008 Pearson Education, Inc.

Regulation of Blood Glucose

Name _____

Date _____ Section _____

Score/Grade _____

Glucose Tolerance Test

1. Record the blood and urine glucose data for the subject from your team and the average values for all team subjects in the class. Plot the blood glucose data on the following graph.

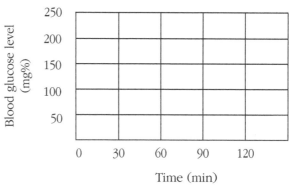

Time (min)

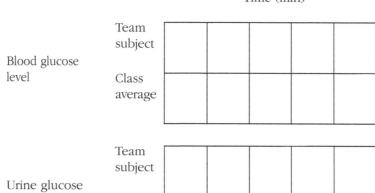

Blood glucose level

Team subject

Class average

Urine glucose level

Team subject

Class average

2. List the effect of each of the following hormones on blood glucose and the mechanism producing the effect.

Hormone	Blood Glucose Effect	Mechanism
Insulin		
Epinephrine		
Glucagon		
Growth hormone		
Cortisol		

Copyright © 2015, 2011, 2008 Pearson Education, Inc.

3. How are the levels of insulin and glucagon regulated in the body?

_____

_____

4. What causes the insulin shock seen when an overdose of insulin is given to an organism?

_____

_____

5. Why does a person who has diabetes mellitus have more acidic urine?

_____

_____

6. Some diabetics control their blood glucose level by ingesting tablets rather than by receiving injections of insulin. How do these tablets work, and who may use them?

_____

_____

7. Define the following terms:

Glycogenolysis:

_____

_____

Gluconeogenesis:

_____

_____

Ketonemia:

_____

_____

Hyperglycemia:

_____

_____

Glycogen:

_____

_____

Glycosuria:

_____

_____

APPLY WHAT YOU KNOW

1. What causes an increase in urine output (diuresis) in diabetes mellitus?

_____

_____

2. The glycemic index (GI) measures the rate at which carbohydrates are absorbed into the bloodstream. Those foods with a high glycemic index, such as sugar, are absorbed very fast, and those with a low glycemic index, such as lentils, are absorbed very slowly. Would the preferred food for a diabetic be one with a high GI or a low GI? Explain your answer.

_____

_____

Copyright © 2015, 2011, 2008 Pearson Education, Inc.

Measurement of Metabolic Rate

CHAPTER 13 INCLUDES:

PhysioEx™ 9.1   For more exercises on Measurement of Metabolic Rate, visit PhysioEx™
(www.physioex.com) and choose Exercise 4: Endocrine System Physiology.

⚠ CAUTION!

Because these experiments involve the use of living vertebrates, it is important that you obtain the appropriate permissions from your Institutional Animal Care and Use Committee (IACUC). While the instructions given are suitable to obtain the desired result, your IACUC may require a modification of procedures to meet your institution's needs.

OBJECTIVES

After completing this exercise, you should be able to

1. Identify the difference between direct and indirect measures of human metabolism.
2. Explain the constraints under which basal metabolic rate is measured.
3. Examine the effects of hormones on metabolic rates.
4. Measure human metabolism, using oxygen consumption rates.
5. Understand the relationship of metabolism to surface area and body weight by comparing human and rat metabolic rates.

Human Metabolism

Metabolism is a broad term that refers to all of the chemical reactions that occur in a biological system, that is, all the reactions of **anabolism** (synthesis of complex molecules from simple molecules) and **catabolism** (breakdown of complex molecules to simple molecules and release of energy). A measurement of metabolism provides information on how the organism obtains its energy and how quickly and efficiently it uses it. **Metabolic rate** measures the amount of energy used by a person's body per unit of time. To make comparisons between animals of different sizes and weights, the metabolic rate is usually expressed as kilocalories per unit of body weight per hour or kilocalories per square meter of body surface area per hour.

A person's metabolic rate depends upon a number of factors including whether they have eaten or are fasting, the types and proportions of food they are consuming, their activity level, differences in their hormonal levels, the amount of mental stress they are experiencing, and the level of thyroid hormones they have. Typically the metabolic rate will rise due to physical activity, anxiety, eating, pregnancy, and experiencing a fever. Starvation and depression can reduce the metabolic rate. Monitoring substantial changes in the metabolic rate of an individual may alert a physician to potential problems in a patient.

Human Metabolism: Calorimetry

Metabolism would be extremely difficult to measure but for the fact that nearly all the energy the body uses is eventually converted to **heat**. Only external work, such as the lifting of a weight, is not converted to heat. Thus, if we can measure the amount of heat produced by the organism when it is not doing work, we will have a valid measure of its metabolic rate. Such a measurement is called **calorimetry** and is usually expressed as calories or kilocalories of heat produced.

Direct Calorimetry

At first glance, it would appear that the easiest way to measure the heat production (metabolism) of an organism would be to measure directly the amount of heat evolved from the body over a certain period of time. This measurement can be done in a special instrument called a **calorimeter**, an insulated chamber containing a water jacket that absorbs the evolved body heat with consequent

Copyright © 2015, 2011, 2008 Pearson Education, Inc.

increase in the temperature of the water. From the amount of water in the jacket and the temperature rise, we can calculate the heat evolved, using the fact that 1 cal is the amount of heat needed to raise 1 g of water 1°C. However, this direct method is quite tedious and difficult and has been largely replaced by simpler indirect methods.

Indirect Calorimetry

The indirect method uses the fact that various food-stuffs (proteins, fats, carbohydrates) will produce nearly the same amount of heat whether they are burned in the body (*in vivo*) or outside the body (*in vitro*). Also, the same amount of oxygen must be used both *in vivo* and *in vitro* for the complete oxidation of these foodstuffs. The heat of combustion and the oxygen consumed when 1 g of each foodstuff is metabolized are shown in the Table 13.1. From the kilocalories per gram and the liters of oxygen per gram, we can calculate the **caloric equivalent of oxygen** in kilocalories per liter of oxygen consumed. Thus, we can determine the metabolic rate (heat production) indirectly by measuring the oxygen consumption of the organism. This is a much simpler method than the direct calorimetry method. For a person on an average balanced diet of carbohydrates, fats, and proteins, we use an average figure of 4.825 kcal/L O_2 consumed.

Basal Metabolic Rate

The basal metabolic rate (BMR) is a measure of the rate of energy use in an "awake" subject in the resting state. It is a measure of the minimal amount of energy needed to maintain just the vital vegetative processes of the body. To be valid, the BMR must be measured under the following rigid conditions:

1. The test is conducted 12 hr or more after the last food is eaten.
2. It is taken after a restful night's sleep, when the activity of the sympathetic nervous system is at its lowest.
3. The subject has been awake and at rest for 30–60 min prior to the test, with no exercise during this period.

4. The subject is in a reclining position in a quiet room whose temperature is between 62° and 87°F.

Obviously, these conditions for an exact BMR cannot be met in the laboratory; therefore, the values obtained in a lab should more correctly be labeled metabolic rate (under the existing conditions) rather than basal metabolic rate. The human BMR is commonly expressed as a percentage of the normal standard metabolic rate as found in tables based on age, sex, and body surface area. A BMR of ± 10% is considered normal.

▶ ACTIVITY 13.1
Human Metabolism
Materials (per group)

☐ Sanborn respirometer

☐ Oxygen tank with regulator

☐ Recording paper

Human metabolism is measured using a **respirometer** or **metabolator** (Figure 13.1). This instrument resembles a spirometer but is modified so that it can be filled with 100% oxygen, and it contains a canister of soda lime for the absorption of carbon dioxide. As the subject breathes in and out of this closed system, oxygen is used from the tank, and the expired carbon dioxide is immediately absorbed by the soda lime. Thus, any decrease in the tank volume represents the volume of oxygen consumed by the subject.

1. Calculation of Oxygen Consumption

a. Let the subject rest in a reclining position for 10–15 min. During this time, fill the respirometer with 100% oxygen and adjust the instrument for recording.

b. During the last 5 min of rest, connect the respirometer to the subject, using a clean mouthpiece and a nose clamp on the nostrils. Adjust the valve near the mouthpiece so that the subject is breathing atmospheric air.

TABLE 13.1   Metabolic Values for Foodstuffs

	Carbohydrates	Fats	Proteins
Heat of combustion (kcal/g)	4.1	9.3	4.3
Oxygen consumed (L O_2/g)	0.75	2.03	0.97
Caloric equivalent of oxygen (kcal/L O_2)	5.0	4.7	4.5

Copyright © 2015, 2011, 2008 Pearson Education, Inc.

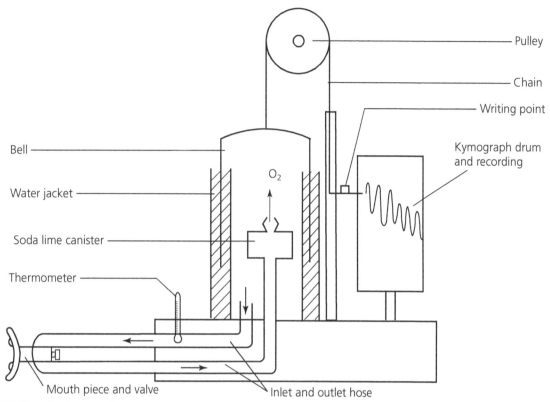

FIGURE 13.1 Sanborn respirometer.

c. After the subject has breathed room air for 5 min, close the mouthpiece valve. Let the subject breathe pure oxygen as naturally as possible for 9–10 min. The first 1–2 min should not be used in your calculations because the respiratory pattern is abnormal while the body is adjusting to breathing pure oxygen. The normal record will resemble that shown in Figure 13.2.

d. As oxygen is used from the tank, the tank drops and the writing pen moves up on the recording paper. The distance between each numbered vertical line on the record shown in Figure 13.2 represents 1 min of time. Draw an average slope line by connecting the expiratory excursions as shown

in the figure. Draw a horizontal line from the beginning of the second or third minute. In Figure 13.2, the distance X represents the distance the tank drops in 4 min. Measure the tank drop in millimeters and multiply this distance by the **tank constant** (ml/mm) for your respirometer. (The tank constant is stamped on the respirometer.) Dividing the resulting volume by the minutes of respiration during this tank drop gives the gross oxygen consumption in milliliters per minute. Place this value on your metabolism data sheet and attach the record to the Laboratory Report. With some respirometers, the oxygen consumption can be obtained directly from the recording paper.

2. Calculation of Metabolic Rate

The metabolism data sheet is arranged so that the metabolic rate can be calculated in an orderly, stepwise manner. This sheet is for your convenience, but you are cautioned not to make your calculations by rote; rather, try to understand why each step is performed. Use the following procedure to calculate metabolic rate.

a. Determine gross oxygen consumption (from your record).

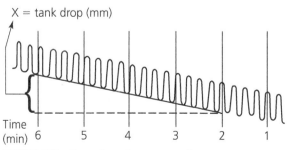

FIGURE 13.2 Normal respiratory record.

b. Determine the room barometric pressure (from a barometer) and the temperature of the oxygen during the run (from the thermometer on the respirometer).

c. Calculate the **STPD (standard temperature pressure dry) factor**. This factor is used to convert the volume of oxygen consumed under the experimental conditions to the volume it would occupy as a dry gas at standard temperature ($0°C$ or 273 K) and standard pressure (760 mm Hg).

$$ \text{STPD} = \frac{\begin{bmatrix} \text{Room} \\ \text{barometric} \\ \text{pressure} \\ \text{(mm Hg)} \end{bmatrix} - \begin{bmatrix} \text{Water} \\ \text{vapor} \\ \text{pressure} \\ \text{(mm Hg)} \end{bmatrix}}{\begin{matrix} \text{Standard pressure} \\ \text{(760 mm Hg)} \end{matrix}} \times \frac{\text{Standard} \atop \text{temp.} \atop (273\ K)}{273\ K\ + \atop O_2\ \text{temp.} \atop (°C)} $$

Obtain the vapor pressure of water at the oxygen temperature from a standard table (Table 2 in Appendix A).

d. Calculate the corrected oxygen consumption volume.

$$ \text{Corrected } O_2 \text{ volume} = \text{Gross } O_2 \text{ volume} \\ \times \text{STPD} $$

e. Calculate the heat produced.

$$ \text{Heat produced (kcal/hr)} = \text{Corrected } O_2 \\ \text{volume (L } O_2/\text{hr)} \times 4.825 \text{ kcal/L } O_2 $$

f. Calculate the metabolic rate (heat produced per unit of surface area).

$$ \text{Metabolic rate (kcal/m}^2/\text{hr)} = \frac{\text{Heat produced (kcal/hr)}}{\text{Body surface area (m}^2)} $$

Obtain the body surface area using the DuBois and DuBois formula.

$$ \text{Body surface area (m}^2) = 0.20247 \times \text{height}^{0.725} \text{ (m)} \\ \times \text{weight}^{0.425} \text{ (kg)} $$

g. Calculate the BMR.

$$ \text{BMR (\%)} = \frac{\begin{bmatrix} \text{Measured} \\ \text{metabolic} \\ \text{rate} \end{bmatrix} - \begin{bmatrix} \text{Normal} \\ \text{metabolic} \\ \text{rate} \end{bmatrix}}{\text{Normal metabolic rate}} \times 100 $$

Calculate the normal metabolic rate using the Mifflin-St. Jeor equation.

Men $= \{ [10 \times \text{weight (kg)}] + [6.25 \times \text{height (cm)}] \\ - [5 \times \text{age (y)}] + 5 \} / 24$

Women $= \{ [10 \times \text{weight (kg)}] + [6.25 \times \text{height (cm)}] \\ - [5 \times \text{age (y)}] - 161 \} / 24$

Determine the resting metabolic rate with the subject reclining. If desired, you can obtain the exercise metabolic rate after having the subject run in place for 200 steps or do a similar exercise. **Enter your observations and perform the calculations in the Laboratory Report.** ◀

Relationship of Metabolism to Surface Area and Body Weight

It has long been recognized that metabolism is related to body size, and when different-sized animals are compared, we usually express their metabolism per unit of body weight (kcal/kg) or surface area (kcal/m²). The relationship of metabolism to body weight and surface area is not, however, as simple as was once believed, and it has generated considerable controversy and numerous research studies in the past 150 years.

● STOP AND THINK

Because surface area to volume ratio is an important factor in determining loss of body heat, it is not surprising that an animal will use a variety of strategies to help conserve body heat, especially if the animal is small. If you have a pet cat or dog, you might have observed them using some obvious behavioral adaptations to minimize heat loss. Describe some of these behaviors and identify how your pet changes behavior to modify its surface area to volume ratio, either alone or by using another animal (including you!).

Perhaps the most famous generalization that came from these studies was the **surface area law**. Proponents of this law noted that the rate of heat loss of a body is proportional to its surface area, and, because heat production equals heat loss in resting animals, they argued that heat production must also be proportional to surface area. Surface area is roughly proportional to weight$^{0.67}$ for objects of similar geometry and specific gravity. The DuBois studies were based on the surface area law. Although the kcal/m² relationship has been useful as an empirical approximation, it is not founded on strong physiological concepts. The surface area value of an animal changes with changes in body position, heat loss varies greatly over various parts of the body, and heat loss and metabolic rate are governed by complex neural and hormonal mechanisms. An added complication is that it is almost impossible to measure the surface area of animals accurately.

The relationship of metabolism to body weight also seemed obvious to early investigators, but it was soon evident that this is not a simple linear relationship. On a kilogram basis, the metabolic rate was found to be higher in small animals than in large animals; hence, the metabolic rate is actually related to a power of body weight, as expressed in the following formula:

$$ M = aW^b $$
$$ \log M = \log a + b \log W $$

Copyright © 2015, 2011, 2008 Pearson Education, Inc.

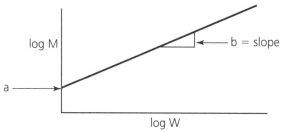

FIGURE 13.3   Plot of log of metabolic rate against log of body weight.

where M = metabolic rate, W = body weight, a = metabolic rate/unit weight, and b = the rate at which metabolism changes with size.

When the log of metabolic rate is plotted against the log of body weight, a constant linear relationship is found (Figure 13.3); this relationship has been demonstrated for fish, amphibians, reptiles, birds, and mammals. Over the years, the most heated controversy has arisen over the exact value of the exponent b, whose average value is around 0.75 but may vary from 0.7 to 0.8, depending on the species. While it is certainly naïve to expect a regression line to accurately predict the metabolic rate of a particular individual or species, significant deviations from the "normal" metabolic level may be related to heat generation and conservation issues associated with where an animal lives.

⬤ STOP AND THINK

Seals and whales have a metabolic rate about twice as high as that predicted by their size. Similarly, desert mammals as variable as camels and rodents have metabolic rates substantially lower than expected from their body size. How would you explain these discrepancies?

In the following activity, you will determine your own value for b by constructing a simple version of Benedict's famous mouse-to-elephant curve for mammals, which we will call a mouse-to-human curve. You will also relate metabolic rate to body surface area for the three animals (mouse, rat, and human) to examine the validity of the surface area law.

▶ ACTIVITY 13.2
Relation of Metabolism to Surface Area and Body Weight

Materials (per group)

☐ 1 rat or mouse

☐ 1 metabolator set up for rat or mouse (see Figures 13.4, 14.1, or 14.2)

1. **Rate of Oxygen Consumption of Human**
   Determine the rate of oxygen consumption of the human as you did in Activity 13.1 or use the data you obtained then.
2. **Rate of Oxygen Consumption of Rat**
   Determine the rate of oxygen consumption of the rat by using the technique described in Chapter 14 (Thyroid Function).
3. **Rate of Oxygen Consumption of Mouse**
   To measure oxygen consumption of the mouse, use either a smaller desiccator than you used for the rat (Figures 14.1 or 14.2) or make a simple mouse metabolator (Figure 13.4).

   a. Determine the oxygen consumption of the mouse by following the steps that you used to determine it for the rat.
   b. Convert the volume of oxygen consumption to the volume of oxygen under standard

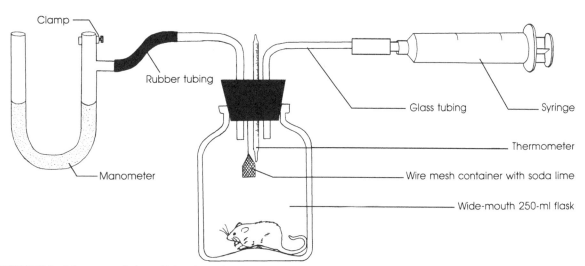

FIGURE 13.4   Mouse metabolator. Using the syringe to maintain the levels of liquid in the manometer will allow for measurements of oxygen consumption over time.

Copyright © 2015, 2011, 2008 Pearson Education, Inc.

conditions (STPD) and calculate the heat produced (in kilocalories per hour and per day), using the conversion factor of 4.825 kcal/L O_2.

c. Express the metabolic rate as kilocalories per square meter per hour. Use the DuBois and DuBois formula and Table 3 in Appendix A to ascertain the body surface areas of the human and the rat, respectively. Calculate the surface area of the mouse from the following equation:

Surface area (m^2) = Weight$^{0.425}$(kg) × Length$^{0.725}$(cm) × 0.007184

log Surface area (m^2) = 0.425 log [Weight (kg)] + 0.725 log [Length (cm)] + log 0.007184

The length of the mouse is measured from the tip of the nose to the base of the tail.

4. **Plotting the Mouse-to-Human Curve**
Using the 3 × 3- or 4 × 4-cycle logarithmic graph paper at the end of the Laboratory Report, plot the metabolism (kcal/day) against the body weight (kg) and draw a mouse-to-human curve. From this plot, determine the a and b values for the formula $M = aW^b$. **Enter your observations and perform the calculations in the Laboratory Report.** ◄

Hormonal Regulation of Metabolism

Chapter 12 illustrates how glucose regulation in the body takes place through the combined actions of insulin and glucagon. However, the effects of glucagon on increasing blood sugar levels when needed would be minimized in the absence of the adrenal cortex hormone cortisol. Cortisol is just one of several hormones that can alter metabolic activity in the body.

▶ ACTIVITY 13.3
Hormonal Regulation of Metabolism

Materials (per student)

☐ Physiology textbook

In this activity, you will examine the effects of several hormones on metabolic rates. Use your textbook to research the effects of insulin, glucagon, epinephrine, growth hormone, cortisol, thyroxine, parathyroid hormone, and calcitonin on human metabolic activity. **Enter this information in your Laboratory Report.** ◄

Copyright © 2015, 2011, 2008 Pearson Education, Inc.

Measurement of Metabolic Rate

Name _____

Date _____ Section _____

Score/Grade _____

Human Metabolism: Calorimetry

Human Metabolism Data Sheet

Name _____ Age _____ Sex _____

Height (cm) _____ Weight (kg) _____ Surface area (m^2) _____

Step	Sample Calculation for a 70-kg Male, 180 cm Tall, Age 20	Experimental Subject
1. Gross oxygen consumption (ml/min)	400	
2. Barometric pressure (mm Hg)	700	
3. Oxygen temperature (°C)	27	
4. Vapor pressure of H_2O (mm Hg)	26.7	
5. STPD factor	0.806	
6. Corrected oxygen consumption (ml/min)	322.4	
7. Oxygen consumption (ml/hr)	19344	
8. Oxygen consumption (L/hr)	19.344	
9. Heat produced (kcal/hr)	93.33	
10. Metabolic rate (kcal/m^2/hr)	49.38	
11. Normal metabolic rate (kcal/m^2/hr)	41.43	
12. BMR (%)	+19.19%	

Respirometer record used in calculating your oxygen consumption (place your record in the following space).

1. What is the purpose of correcting the oxygen volume by using the STPD factor?

_____

_____

Copyright © 2015, 2011, 2008 Pearson Education, Inc.

2. Why is metabolism expressed in relation to body surface area?

_____

_____

3. List five factors that alter metabolic rate and how each affects it.

_____

_____

_____

4. Assume that your resting gross carbon dioxide production for 5 min is 1.6 L. Calculate your resting respiratory quotient (RQ) (RQ = CO_2 produced/O_2 consumed). RQ = _____
   What is the importance of the RQ in physiology?

_____

_____

Relationship of Metabolism to Surface Area

Measurement	Mouse	Rat	Human
Weight (kg)			
Surface area (m^2)			
Gross O_2 consumption (L/hr)			
Corrected O_2 consumption (L/hr)			
Heat produced (kcal/hr)			
Heat produced (kcal/day)			
Metabolic rate (kcal/m^2/hr)			

1. What is the theoretical basis for relating metabolism to surface area?

_____

_____

2. Are the metabolic rates of the three mammals comparable when based on square meters of body surface area?

_____

_____

Relationship of Metabolism to Body Weight

1. Calculate the metabolic rate of the three animals in terms of body weight (kcal/kg/hr).

	Mouse	Rat	Human
Metabolic rate (kcal/kg/hr)			

Attach your log–log plot of metabolism (kcal/day) versus body weight (kg) to the lab report.

Copyright © 2015, 2011, 2008 Pearson Education, Inc.

2. What exponent *b* value did you obtain? _____ What does this *b* value tell you about the relationship between metabolism and body weight?

_____

_____

3. How does your exponent *b* compare with the values, usually cited in the literature, of around 0.65–0.75?

_____

_____

4. Which of the three means of expressing metabolic rate do you think allows the best comparison of different-sized animals? Why?

_____

_____

Hormone Regulation of Metabolism

The capture and consumption of energy by the body is regulated by the action of several hormones that promote anabolism or catabolism of the body's nutrients and minerals. List the major effects of the following hormones on nutrients and minerals, using (+) for stimulatory and (−) for inhibitory actions. Also include any other physiological effects associated with these hormones that are unrelated to their metabolic actions.

Insulin:

_____

_____

Glucagon:

_____

_____

Epinephrine:

_____

_____

Growth hormone:

_____

_____

Cortisol:

_____

_____

Thyroxine:

_____

_____

Parathyroid hormone:

_____

_____

Calcitonin:

_____

_____

Copyright © 2015, 2011, 2008 Pearson Education, Inc.

APPLY WHAT YOU KNOW

1. Epinephrine and glucocorticoid hormones are released in response to stress. Use your knowledge of their functions to explain the significance of these secretions during stressful periods.

_____

_____

_____

2. Is the determination of metabolic rate a valid test for hyperthyroidism or hypothyroidism? Why or why not?

_____

_____

_____

For Activity 13.2.4: Plotting the Mouse-to-Human Curve

Copyright © 2015, 2011, 2008 Pearson Education, Inc.

Thyroid Function

CHAPTER 14 INCLUDES:

Vernier 1 Vernier® Activity

PhysioEx™ 9.I For more exercises on Thyroid Function, visit PhysioEx™ (www.physioex.com) and choose Exercise 4: Endocrine System Physiology.

⚠ CAUTION!

Because these experiments involve the use of living vertebrates, it is important that you obtain the appropriate permissions from your Institutional Animal Care and Use Committee (IACUC). While the instructions given are suitable to obtain the desired result, your IACUC may require a modification of procedures to meet your institution's needs.

OBJECTIVES

After completing this exercise, you should be able to

1. Examine the effects of thyroxine on metabolism.

Thyroid Effects on Metabolism

The thyroid is a shieldlike gland located on the ventral side of the trachea. Under the influence of **thyrotropin**, or thyroid-stimulating hormone (TSH), from the anterior pituitary, the thyroid secretes several thyroid hormones, the chief of which is **thyroxine**. Thyroxine is a powerful hormone that affects the metabolism of every body cell; its principal effects are an increase in oxygen consumption and calorigenesis (heat production) by the cells. The exact mechanism of action of thyroxine is still unknown. The general effects of thyroxine can be observed in the hyperthyroid person. He or she displays an increase in heart rate, stroke volume, and blood pressure and increased peripheral vasodilation to aid loss of excess body heat. The cells of the nervous system become hyperexcitable so that the person becomes irritable and nervous. The increased metabolic rate depletes the body's fat reserves and the person becomes quite thin. The hypothyroid person shows opposite symptoms: increased body weight, sluggishness, low metabolism, and low heat production.

The metabolic effects of thyroxine can be observed in animals fed thyroxine in excess for a period of 2–3 weeks. A latent period of about 10 days occurs before the effects of excess or deficient thyroxine can be seen. In the following activity, you will examine thyroxine's actions by measuring the metabolic rate of normal, hypothyroid, and hyperthyroid rats. Hypothyroidism can be induced by adding thiouracil to the diet. A thiouracil-derived drug, Propylthiouracil, is used to treat hyperthyroidism and acts by inhibiting the enzyme thyroperoxidase, reducing the production of thyroxine. The hyperthyroid condition is induced by injecting a rat with triiodothyronine or L-thyronine.

▶ ACTIVITY 14.1
Metabolic Effects of Thyroxine

Materials (per three groups)

☐ 6 rats (either all male or all female) 150–200 g in weight

1. Preparation of Animals

The control experimental animals should be of the same sex and should be of young (150–200 g) stock not used for other experiments. Two weeks prior to the laboratory period, the rats will be divided into the following three groups (two rats per group):

a. Euthyroid controls.
b. Hyperthyroid. Fed 1% desiccated thyroid in their rat chow for 2 weeks. An alternate technique is to give the rats an intraperitoneal (IP) injection of 15 µg of triiodothyronine or 25 µg of L-thyronine every third day.

Copyright © 2015, 2011, 2008 Pearson Education, Inc.

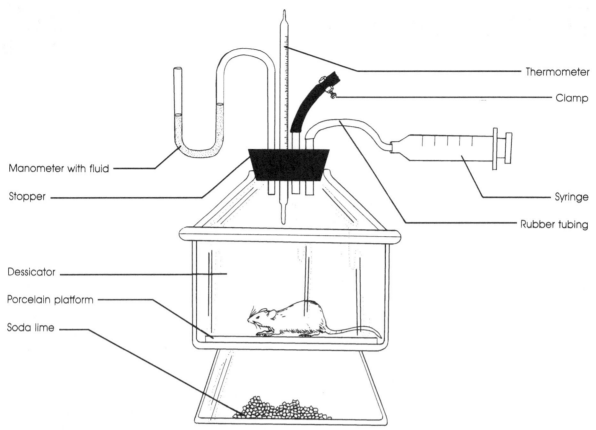

FIGURE 14.1   Apparatus for determining the metabolic rates of small animals.

c. Hypothyroid. Fed 0.5% thiouracil or 0.02% propylthiouracil in their rat chow for 2 weeks.

The weights of all rats should be recorded weekly.

2. Principles

The animal is placed in a closed system in which carbon dioxide is continuously removed by reaction with soda lime (Figure 14.1). The pressure in the respirometer is kept constant so that any change in volume represents oxygen consumed by the animal. The volume change is measured by means of a Vaseline-coated syringe. The heat production of the animal is calculated on the assumption that for each liter of oxygen consumed, 4.825 kcal of heat is produced. This constant is called the **caloric equivalent of oxygen**.

Oxygen consumed can also be measured using an oxygen gas sensor and the Vernier software. The sensor uses an electrochemical cell to measure oxygen concentrations from 0 to 27%. Because this system measures a change in the percentage concentration of oxygen in a closed system, it is important to know the volume of the dessicator (including the cover) being used.

3. Procedure

Materials (per group)

☐ 1 rat or mouse

☐ 1 metabolator set up for rat or mouse (see Figures 13.4 or 14.1)

You will be assigned animals from one of the three experimental groups. Measure the metabolic rate of your rats before and after the 2-week feeding period. Weigh the rats and obtain their body surface area (using Table 3 in Appendix A).

When measuring oxygen consumption using a desiccator, it is essential to follow the directions exactly. The following precautions will help ensure the success of your measurements:

1. Always keep the clamp on the vertical tube open while making the initial adjustments.
2. Never push abruptly on the syringe, or you will separate the manometer fluid column.
3. During the run, keep the manometer fluid level at all times; this keeps the pressure in the chamber constant.
4. Handle the apparatus and the rat gently at all times.

Copyright © 2015, 2011, 2008 Pearson Education, Inc.

Each team will obtain a rat and determine oxygen consumption as follows:

a. Weigh your rat to the nearest gram and record the weight.
b. Remove the top half of the desiccator and place the rat inside. Make certain that adequate fresh soda lime is in the bottom of the desiccator. Be sure the clamp on the vertical tube is open.
c. Replace the top of the desiccator, using Vaseline to provide an airtight seal between the two halves.
d. Test the system for leaks by closing the clamp on the vertical tube and gently pulling back on the plunger of the syringe. The manometer fluid should move toward the desiccator. If there is a leak in the system, the manometer fluid will return to its original level. If you detect a leak, try to seal it by applying Vaseline around the stopper or wherever there is a rubber-to-glass connection.
e. Open the clamp on the vertical tube and adjust the card on the back of the manometer so that the crossline is level with both menisci of the manometer fluid.
f. Wait 4–5 min to allow the temperature to reach equilibrium in the respirator.
g. Close the clamp on the vertical tube and record the time.
h. Manipulate the plunger of the syringe to keep the two limbs of the manometer fluid constantly on the crossmark.
i. Exactly every minute (or every 2 min), read the syringe and record the reading. Also read the thermometer and record the temperature.
j. Continue making observations until the syringe is empty.
k. Open the vertical tube clamp and remove the top half of the desiccator to allow the stale air to be flushed out.
l. Repeat the previous steps two or three times and determine the average milliliters of oxygen used per minute by your rat.

4. Calculation of Metabolic Rate
Using the metabolism data sheet in the Laboratory Report as a guide, calculate the oxygen consumption as milliliters of oxygen per minute per 100 g body weight and the heat production as kilocalories per square meter per hour. It is necessary to correct the observed gas volume to the volume it would occupy under standard conditions of pressure and temperature (pressure = 760 mm Hg; temperature = 0°C or 273 K). To make this

correction, multiply the observed milliliters of oxygen per minute consumed by the STPD factor (see Chapter 13). Summarize the data from all the rats in the lab to see the effects of the thyroid on metabolism and record the average values in the table in the Laboratory Report.

14.1 VERNIER® VERSION

Materials (per group)

☐ 1 rat or mouse
☐ 1 desiccator
☐ 1 Vernier Oxygen meter and cables
☐ 1 computer
☐ Lab Pro software

Materials needed for this exercise will have been set up for you (and are shown in Figure 14.2). Prepare the computer to monitor oxygen levels by opening the file in the Experiment 11a folder of *Biology with Computers*. A graph with oxygen concentration (in %) on the vertical axis and time on the horizontal axis will appear. Make sure that you have followed the instructions to calibrate the oxygen probe if this is necessary.

Click **Data Collection**. Click the **Collection** tab. A window will open. Set **Mode** to **Time Based**. Enter **10** and choose minutes from the drop-down menu. Select **Sample at Time Zero**, set sampling rate to **Medium**, and set six samples to be collected per minute. These settings will instruct the computer to collect data for 10 min at a rate of six samples per minute.

1. Weigh your rat to the nearest gram and record the weight.
2. Remove the top half of the desiccator, keep it off for 5 min to allow air to circulate, and record the temperature of the air in the dessicator. No soda lime is required in the dessicator. The volume of the dessicator (in ml) should be marked on it.
3. Place the rat inside the dessicator. Replace the top half of the dessicator, using Vaseline to provide an airtight seal between the two halves.
4. Click **Collect** immediately to begin making measurements. You will notice a live reading of the dessicator's oxygen concentration in the bottom left of the screen. Note the beginning oxygen concentration. Stop the experiment before the 10-min period if this value goes below 14%.

Copyright © 2015, 2011, 2008 Pearson Education, Inc.

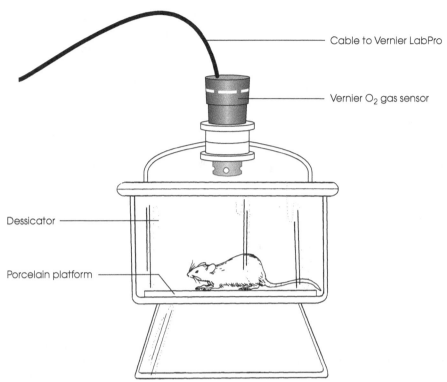

Cable to Vernier LabPro

Vernier O$_2$ gas sensor

Dessicator

Porcelain platform

FIGURE 14.2 Apparatus for determining the metabolic rates of small animals using the Vernier® oxygen gas sensor.

5. Data will be collected for a 10-min period and will automatically be graphed on screen and recorded in the table. Note the ending concentration.

6. Make sure that you return the rat to the cage after the 10-min period.

7. The volume of oxygen consumed can be calculated using the following method.

Beginning oxygen concentration = _____ %

Ending oxygen concentration = _____ %

Difference (D) = _____ %

Volume of oxygen consumed (V) =

$$\frac{(\text{Dessicator volume}) \times D}{100}$$

Length of experiment (T) = _____ min

Gross oxygen consumed (ml/min) = $\dfrac{V}{T}$ =

_____ ml/min

Enter this value in step 5 of the metabolism data sheet in the Laboratory Report and perform all calculations to estimate metabolic rates. ◄

Copyright © 2015, 2011, 2008 Pearson Education, Inc.

Thyroid Function

Name _____

Date _____ Section _____

Score/Grade _____

Metabolism Data Sheet (Rats)

Animal number or marking _____ Weight (g) _____

Treatment _____ Surface area (m²)* _____

Volume of desiccator _____

Step	Example Calculation for 200-g Rat	Experimental Animals	
1. Chamber temperature (°C)	28		
2. Barometric pressure (mm Hg)	740		
3. Vapor pressure of H_2O (mm Hg)	28.3		
4. STPD (standard temperature, pressure, dry) factor	0.845		
5. Average gross oxygen consumption (ml/min)	5		
6. Corrected oxygen consumption (gross × STPD) (ml/min)	4.225		
7. Oxygen consumption per unit body weight (ml/[100g body wt-min])	2.1125		
8. Heat production (corrected oxygen consumption × 0.004825) (kcal/min)	0.0203856		
9. Heat production (kcal/hr)	1.223136		
10. Metabolic rate per unit surface area (kcal/m²-hr)	40.63		

(*See Table 3 in Appendix A.)

Thyroid Effects on Metabolism

1. Record the metabolic rates (kcal/m²-hr) for the rats in each experimental group, both pretreatment (before feeding) and post-treatment (after feeding). Use the average values for rats in each group.

	Metabolic Rate (kcal/m²-hr)		
	Euthyroid (Control)	Hyperthyroid (Thyroid)	Hypothyroid (Thiouracil)
Pretreatment			
Post-treatment			
Difference between pretreatment and post-treatment			
Percent change			

Copyright © 2015, 2011, 2008 Pearson Education, Inc.

2. What differences do you observe between each group of rats when measuring their pretreatment and post-treatment metabolic rates?

_____

_____

_____

3. Explain these results on the basis of changes in thyroid activity.

_____

_____

_____

4. What is a goiter? How is a goiter in hyperthyroidism compared with one in hypothyroidism?

_____

_____

_____

5. What role do iodine and the amino acid tyrosine together play in the function of the thyroid gland?

_____

_____

_____

APPLY WHAT YOU KNOW

1. Graves' hyperthyroidism is a disease caused by the development of antibodies that stimulate the TSH receptor. Explain why the symptoms include (a) the development of a goiter, (b) an increase in thyroxine secretion, and (c) low plasma TSH levels.

_____

_____

_____

Copyright © 2015, 2011, 2008 Pearson Education, Inc.

Nerve–Muscle Activity

15

CHAPTER 15 INCLUDES:

 PowerLab 1 PowerLab® Activity

 **Vernier** 1 Vernier® Activity

BIOPAC
Systems, Inc. 1 BIOPAC® Activity

PhysioEx™ 9.1

For more exercises on Nerve–Muscle Activity, visit PhysioEx™ (www.physioex.com) and choose Exercise 2: Skeletal Muscle Physiology and Exercise 3: Neurophysiology and Nerve Impulses.

⚠ CAUTION!

Because these experiments involve the use of living vertebrates, it is important that you obtain the appropriate permissions from your Institutional Animal Care and Use Committee (IACUC). While the instructions given are suitable to obtain the desired result, your IACUC may require a modification of procedures to meet your institution's needs.

OBJECTIVES

After completing this exercise, you should be able to

1. Determine and record the threshold voltage for frog muscle contraction when stimulated directly and through the sciatic nerve.
2. Use the *in vitro* or *in vivo* frog muscle preparation to observe and record the effects of stimulus strength, stimulus frequency, and load on muscle contractility.
3. Set up and observe the response of frog muscle to a tetanizing stimulus to the point of fatigue.
4. Set up and record the effects of tubocurarine on the neuromuscular junction.
5. Determine the response of human muscle to electrical stimulus.
6. Graphically examine the response of human muscle under changing strengths and durations of contraction.

In the body, skeletal muscle contractility is controlled by the conduction of action potentials over motor neurons and the subsequent release of acetylcholine at the neuromuscular junction. Each motor neuron innervates (makes synapse with) several individual muscle fibers (cells). One motor neuron and all the muscle fibers it innervates is called a *motor unit*. The overall contractile strength exerted by a muscle is thus determined by the number of motor units activated by the motor cortex in the brain and the frequency with which these cells are stimulated. As we increase the strength of contraction of a muscle, we recruit additional motor units; the result is graded contractions. Recruitment is not a random process in that the first axons that fire are those serving the smallest motor units. Consequently, contraction strength is relatively low. When excitatory input increases, larger axons are recruited, resulting in larger motor units contracting and an increased strength of contraction. It is believed that we cannot voluntarily recruit the largest motor units; however, during times of extreme anger or fear, humans exhibit unusual feats of strength because they are able to activate all motor units simultaneously.

In the following activities, you will examine the operation of the neuromuscular junction and the major characteristics of muscle contractility. Much of our knowledge in this area was derived from the classic studies of the frog gastrocnemius muscle conducted during the latter half of the nineteenth century. You will also examine muscle activity using human myography.

Dissection of Nerve–Muscle Preparation

Pithing the Frog

You will kill the frog by using a pithing needle to destroy both the brain and the spinal cord (double pithing). Hold the frog in one hand, using your

Copyright © 2015, 2011, 2008 Pearson Education, Inc.

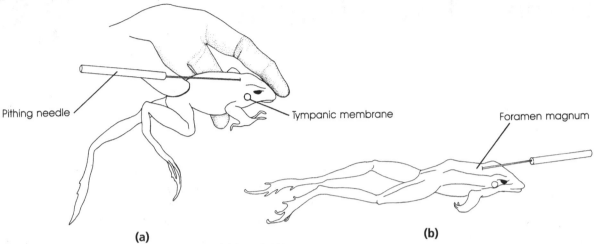

FIGURE 15.1   Pithing of frog. (a) Brain pithing. (b) Spinal pithing.

index finger to bend the head slightly downward. Insert the pithing needle into the slight depression between the skull and spinal cord. This opening (the foramen magnum) is approximately even with the posterior edge of the tympanic membranes.

Insert the needle quickly into the foramen, move the needle to a position parallel to the top of the skull, push the point forward into the brain, and move the point from side to side to destroy the brain (Figure 15.1a). When the brain is successfully pithed, the corneal reflex is lost. Touch the cornea lightly to see if an eye blink can be elicited. Repeat the brain pithing until the corneal reflex is abolished.

A single-pithed frog does not feel pain but still exhibits spinal reflexes. These reflexes can be destroyed by inserting the needle in a foramen and running it down the spinal cord to destroy the cord (Figure 15.1b). When this double pith is complete, the hind legs will become extended and then will relax.

In Vitro Muscle Preparation

Double pith the frog. Then use bone cutters or heavy scissors to cut off a hind leg at the upper end of the thigh. Save the rest of the animal wrapped in a wet paper towel so that the other leg may be prepared if the first muscle fatigues. Use scissors to cut the skin around the ankle and peel back the loose skin from the muscle (Figure 15.2a). Avoid touching the nerve or muscle with dry fingers because this will cause depolarization of the membranes and thereby stimulate spontaneous muscle contractions. Cut the skin on the back of the thigh and, using a glass probe, isolate the

sciatic nerve from the surrounding muscles in the thigh. Cut away the thigh muscles from the femur, leaving the nerve exposed. Pass a thread around the Achilles tendon at the heel and tie it tightly around the tendon. Sever the tendon below the tie and pull the muscle away from the leg. Cut away the fibiotibula bone just below the knee. This procedure will leave the gastrocnemius muscle attached to the knee, with a section of the femur left to anchor the preparation in subsequent experiments (Figure 15.2b).

Place the femur between the jaws of the femur clamp and tighten the clamp securely. Tie the thread attached to the Achilles tendon to the force transducer in the PowerLab® system or to the electronic myograph transducer used in the Physiograph™ system. Position the muscle so it forms a right angle with the femur and is vertical to the lever arm or transducer (Figure 15.3). You may stimulate the muscle by positioning an electrode holder so that the electrode prongs are held against the muscle or by inserting pin electrodes into each end of the muscle and attaching them to the stimulator. *Keep the muscle and nerve moist with frog Ringer's solution at all times.* Twitching of the muscle when it is not being stimulated indicates that the tissue is drying and needs additional frog Ringer's solution.

In Vivo Muscle Preparation

In this alternative preparation, the gastrocnemius muscle is left attached to the body with its blood supply intact.

Double pith the frog. Cut the skin on the back of the thigh on one leg and gently retract the

Copyright © 2015, 2011, 2008 Pearson Education, Inc.

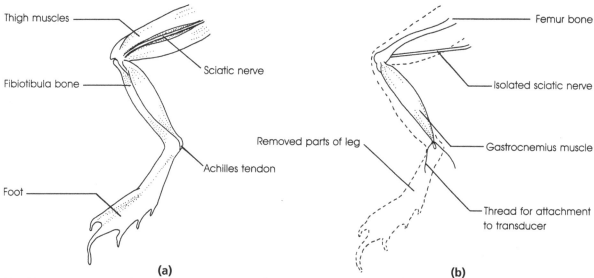

FIGURE 15.2   Dissection of frog gastrocnemius muscle (*in vitro* muscle preparation). (a) Skinned hind leg. (b) Nerve–muscle preparation.

gluteus and semimembranous muscles to expose the sciatic nerve deep between the muscle groups. Using a glass probe, slip a thread under the nerve so it can be lifted up and stimulated with electrodes during later experiments. For now, let the nerve retract between the muscle groups and moisten the area with frog Ringer's solution.

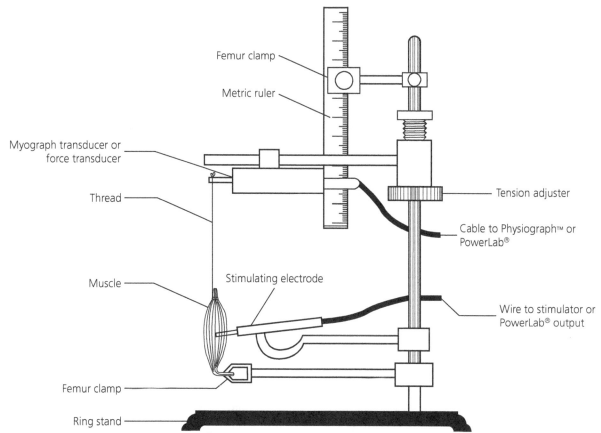

FIGURE 15.3   Physiograph™ or PowerLab® system for recording muscle contractions.

Copyright © 2015, 2011, 2008 Pearson Education, Inc.

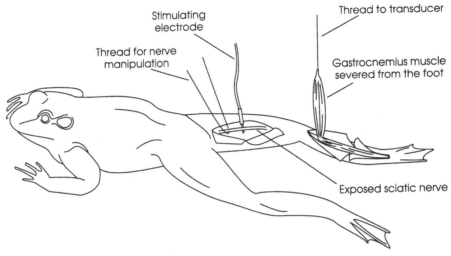

FIGURE 15.4 *In vivo* muscle preparation.

⬤ STOP AND THINK

The contraction of a muscle is often described using the sliding filament theory, in which the shortening of a myofibril is a consequence of an increased overlap between actin and myosin filaments within a sarcomere. However, neuromuscular diseases such as multiple sclerosis, myasthenia gravis, and Parkinson's disease do not result from any problem with muscle fibers. Use your understanding of impulse conduction and synaptic transmission to describe why muscle function is affected in each of these diseases.

Cut and remove the skin over the gastrocnemius muscle. Tie a thread tightly around the Achilles tendon and cut the tendon free of its attachment to the foot. Place the frog prone on the frog board and pin the knee into the holder (Figure 15.4). Tie the free end of the thread to the transducer (as shown in Figure 15.3) and insert pin electrodes into each end of the muscle for stimulation. Use the tension adjuster to apply a slight stretch on the muscle so it is at a better length for contraction. *Keep the muscle moist with frog Ringer's solution at all times.* Cover the rest of the frog with a paper towel soaked in water.

Stimulation of Tissues

If you are unfamiliar with the electronic stimulator, please refer to Chapter 5, "Membrane Action Potentials," the section titled "Stimulation of Tissues," before you begin the experimental procedures. If you are using the PowerLab® system, you will be using the built-in stimulator.

Isolated Muscle Responses

▶ ACTIVITY 15.1

Isolated Muscle Responses (Using the Physiograph™)

Materials

☐ 1 Physiograph™ with recording paper

☐ 1 electronic myograph transducer

☐ 1 set stimulating electrodes

☐ 1 stimulator

☐ 1 frog

☐ 1 femur clamp

☐ 2 glass probes

☐ 1 dissecting needle

☐ 1 6-inch metric ruler

☐ 1 10-cm piece of cotton string

☐ 1 250-ml squeeze bottle of frog Ringer's solution at room temperature

☐ 1 ml of tubocurarine (3 mg/ml)

☐ Dissection equipment (forceps, scissors, bone cutters)

1. Muscle and Nerve Irritability

Double pith the frog and make an *in vivo* preparation of the gastrocnemius muscle, without attaching the muscle to the transducer.

Lift the sciatic nerve out from between the muscles and lay it over the stimulating electrodes. Using a 1-msec pulse duration, stimulate the nerve with a single pulse and determine the lowest voltage (threshold) that will produce a perceptible (visual, not recorded) muscle twitch. Record this voltage.

Copyright © 2015, 2011, 2008 Pearson Education, Inc.

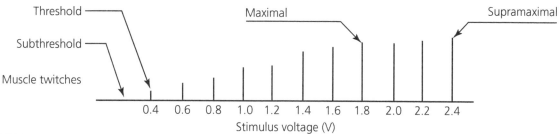

FIGURE 15.5   Muscle contractions at various strengths of stimuli.

Replace the nerve between the muscles and place the stimulating electrodes directly on the gastrocnemius muscle. Determine the threshold voltage that produces a perceptible twitch as was done with the nerve stimulation. Record this voltage.

2. Effects of Stimulus Strength

Prepare the muscle for recording contractions, using the Physiograph™ system. Set the stimulator for 1-msec pulse durations.

Using a very slow paper speed, stimulate the muscle directly with a single pulse and record the first perceptible twitch obtained with a threshold stimulus. Label the recording with the voltage used. Now gradually increase the voltage in a stepwise manner, in small increments, and record the muscle twitches obtained. Continue this process until no further increase in twitch height occurs. Do not exceed this maximal voltage by any large margin or you risk damaging the muscle. Label the record with each voltage applied. You should now be able to categorize stimuli based on the response obtained. Your record should resemble that shown in Figure 15.5.

3. Effects of Stimulus Frequency

Using a moderate paper speed, stimulate the muscle with a supramaximal voltage long enough to obtain five to eight twitches at each of the following frequencies: 0.5/sec, 1/sec, 2/sec, 3/sec, 4/sec, and so on until tetanus is produced. Allow 15 sec for recovery between each series of stimulations and apply frog Ringer's solution to the muscle

during this recovery time. Your record should resemble that shown in Figure 15.6.

What is the effect of the higher frequencies? What happens to the strength of muscle contraction and the extent of muscle relaxation when the frequency of stimulation is increased?

4. Effects of Stretch or Load on Muscle Contractility

The effects of load or stretch on a muscle prior to its contraction were first described by Wallace Fenn in the 1920s. Later, Ernest Starling found that the isolated heart responded in the same manner, and this became widely known as Starling's law of the heart.

This basic characteristic of muscle response can be demonstrated by increasing the weight that a muscle must lift or by stretching the muscle in a progressive, stepwise manner. The method employed in this activity will depend on the recording system available.

Effect of Stretch: Length-Tension Curve: Mount a metric ruler on the ring stand so that muscle length and stretch can be measured by sighting over the rod that holds the transducer (Figure 15.3). Adjust the amplifier sensitivity so that a single supramaximal stimulus voltage produces a pen deflection of around 2 cm. Adjust the pen so it is at the lowest position in the range of pen movement. Then use the tension adjuster to reduce the stretch on the muscle until a supramaximal voltage produces only a very small pen deflection (1–2 mm) when the muscle contracts. Record this as 0 mm of muscle

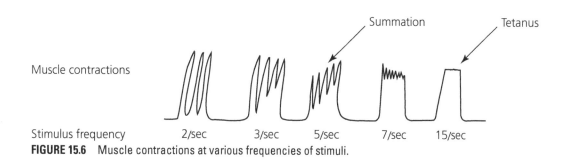

FIGURE 15.6   Muscle contractions at various frequencies of stimuli.

Copyright © 2015, 2011, 2008 Pearson Education, Inc.

stretched length. Record also the ruler measurement at this muscle length by sighting over the rod that holds the transducer.

Stimulate the muscle with a single pulse of supramaximal voltage. Record the millimeters of stretch placed on the muscle by writing under each contraction. Use a very slow paper speed for the rest of the activity.

Stretch the muscle 1 mm and stimulate it as before. Continue in this manner, stretching the muscle in 1 mm increments, stimulating at each new length, and recording the millimeters of stretch under each contraction. Allow at least 10 sec of rest time after each stimulation before the muscle is stretched and stimulated again. Continue the experiment until no contraction occurs when the muscle is stimulated. *Keep the muscle moist with frog Ringer's during the entire experiment.* The record produced should resemble that shown in Figure 15.7.

Note: The baseline of the recording will be progressively elevated as the muscle is passively stretched between active contractions. If the pen deflection exceeds the range of pen movement, reposition the pen lower in the range to provide room for recording the muscle contractions.

From your record, determine the *passive* (resting) tension on the muscle in millimeters of pen deflection from the baseline position. Then determine the *active* (contractile) tension developed when the muscle was stimulated.

In the Laboratory Report, record the millimeters of passive and active tension developed at each muscle stretch length. Then graph the data, with muscle stretch length on the abscissa (*x*-axis) and tension on the ordinate (*y*-axis). Plot all three tensions: active, passive, and total (active + passive) on the same graph, using different colors. At what stretched length does the muscle have the greatest active tension? At what stretched length is the active tension reduced to 0?

5. Neuromuscular Fatigue

Lift the sciatic nerve out from between the muscles and place it over the stimulating electrodes. Stimulate the nerve with a supramaximal voltage and a tetanizing frequency (10–15/sec). Continue stimulating until contraction fatigue is observed (contraction height is reduced by one-half).

Stimulate the nerve with a single shock of supramaximal voltage and record the strength (height) of the muscle contraction. Then place the electrodes directly on the muscle, stimulate it in the same manner, and record the twitch.

Allow the preparation to rest for 5 min and then repeat the nerve and muscle single-shock stimulations. Has complete recovery taken place?

6. Neuromuscular Blockade

Inject 0.1 ml of tubocurarine (3 mg/ml) into the belly of the muscle near the entrance of the nerve. An alternative method is to inject 1 ml of tubocurarine under the skin on the back of the frog so the drug can be absorbed by the lymphatic system into the blood.

Using a very slow paper speed, stimulate the nerve every 10 sec, using the same single supramaximal voltage used previously. Within 5 min, you should see a change in the strength of the muscle contraction.

When the contractions reach one-half the initial height, stop the nerve stimulation. Stimulate the muscle directly once again and record the contraction height.

15.1 POWERLAB® VERSION

Materials

- ☐ 1 computer
- ☐ 1 PowerLab® data acquisition unit
- ☐ 1 force transducer
- ☐ 1 set stimulating electrodes

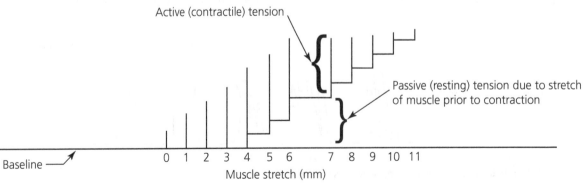

FIGURE 15.7   Contractions of stretched muscle.

Copyright © 2015, 2011, 2008 Pearson Education, Inc.

□ 1 frog

□ 1 femur clamp

□ 2 glass probes

□ 1 dissecting needle

□ 1 6-inch metric ruler

□ 1 10-cm piece of cotton string

□ 1 250-ml squeeze bottle of frog Ringer's solution at room temperature

□ 1 ml of tubocurarine (3 mg/ml)

□ Dissection equipment (forceps, scissors, bone cutters)

Materials needed for this exercise will have been set up for you. You will now set up the Chart software to investigate frog gastrocnemius contractions.

1. Launch **Chart**.
2. Click **Setup** and choose **Channel settings**.
3. In the pop-up window, set channels to **1** and range to **5 mV**.
4. In the same window, click **Input Amplifier**.
5. You will now see a live reading (Figure 15.8) from the force transducer. Set the Low Pass filter to **50 Hz** and select the **AC Coupled**

box. If you tap the transducer, you should observe deflections in the input amplifier window.

6. Click **OK** to close the input amplifier window and once again to close the channel settings window.
7. You will now set up the on-board stimulator. Click **Setup** and choose **Stimulator**. The stimulator window will appear (Figure 15.9).
8. You will make a number of changes in the stimulator window during the course of the activity. The stimulator mode can be set to pulse (single or multiple discrete stimuli are delivered) or step (a series of stimuli is delivered). The options next presented to you will change depending upon whether you chose step or pulse. Each pulse or step can be manipulated to adjust the duration of the pulse/step, when the stimuli are delivered (either manually or after a predetermined delay after recording has started), and the strength (in volts). Click **Close** after you have examined this window in depth. Changes to stimulation frequency, duration, and amplitude can easily be made by

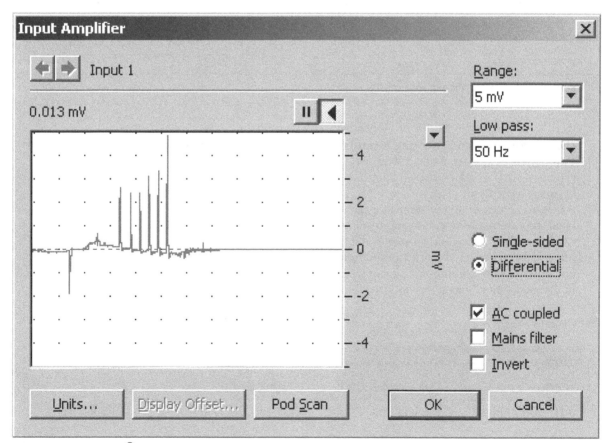

FIGURE 15.8   PowerLab® Input Amplifier window.

Copyright © 2015, 2011, 2008 Pearson Education, Inc.

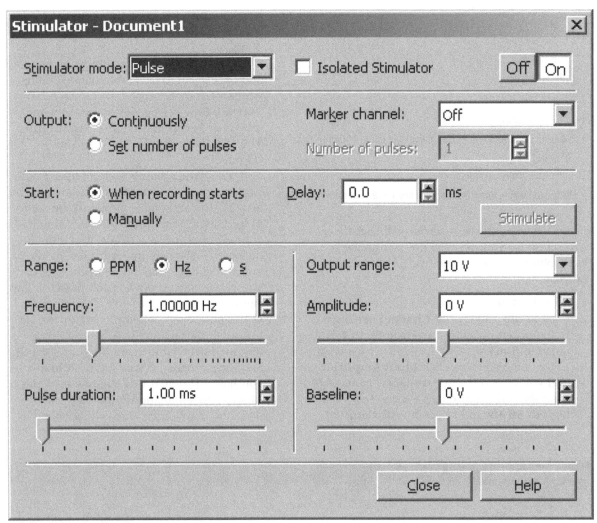

FIGURE 15.9   PowerLab® Stimulator window.

clicking **Settings** and choosing **Stimulator Panel**. This will open a small window (Figure 15.10) in which these changes can be made very easily.

9. When you have set up the stimulator parameters, you are ready to start taking readings. Click **Start** and then click **Stop** after the stimuli have been delivered. Click **Commands** and choose **Add Comment**. You can type in the stimulation voltage or other suitable information in the Add Comment window to label your record.

10. You can print out your record by highlighting the portion you want printed and then clicking **File** and choosing **Print selection**.

1. Muscle and Nerve Irritability

Double pith the frog and make an *in vivo* preparation of the gastrocnemius muscle without attaching the muscle to the transducer.

FIGURE 15.10   PowerLab® Stimulator Panel window.

Copyright © 2015, 2011, 2008 Pearson Education, Inc.

Lift the sciatic nerve out from between the muscles and lay it over the stimulating electrodes.

a. Launch **Chart**.
b. In the stimulator window, set mode to **Step**, in **Output**, select **Set Number Of Pulses**, and set number of pulses to **1**.
c. Set delay to **1000 msec**.

Using a 1-msec pulse duration, stimulate (by clicking **Start**) the nerve with a single pulse and observe the response of the muscle. If no visual response is observed, click **Stop**. Gradually increase the voltage and repeat the procedure until you determine the lowest voltage (threshold) that will produce a perceptible (visual, not recorded) muscle twitch. Record this voltage, using the **Add Comment** window.

Replace the nerve between the muscles and place the stimulating electrodes directly on the gastrocnemius muscle. Determine the threshold voltage that produces a perceptible twitch as was done with the nerve stimulation. Record this voltage.

2. Effects of Stimulus Strength

Prepare the muscle for recording contractions, using the *in vitro* muscle preparation. Set the stimulator for 1-msec pulse durations.

Using a very slow speed (4/sec), stimulate the muscle directly with a single pulse and record the first perceptible twitch obtained with a threshold stimulus. Label the recording with the voltage used. Now gradually increase the voltage in a stepwise manner, in small increments, and record the muscle twitches obtained. Continue this process until no further increase in twitch height occurs. Do not exceed this maximal voltage by any large margin or you risk damaging the muscle. Label the record with each voltage applied. You should now be able to categorize stimuli based on the response obtained. Your record should resemble that shown in Figure 15.5.

3. Effects of Stimulus Frequency

In the Stimulator window, select **Set Number Of Pulses** and set number of pulses to **8**. Set the amplitude to a supramaximal voltage. Using a moderate (10/sec) speed, stimulate the muscle with the supramaximal voltage long enough to obtain eight twitches at each of the following frequencies: 0.5/sec, 1/sec, 2/sec, 3/sec, 4/sec, and so on until tetanus is produced. Allow 15 sec for recovery between each series of stimulations and apply frog Ringer's solution to the muscle during this recovery time. Your record should resemble that shown in Figure 15.6.

What is the effect of the higher frequencies? What happens to the strength of muscle contraction and the extent of muscle relaxation when the frequency of stimulation is increased?

4. Effects of Stretch or Load on Muscle Contractility

The effects of load or stretch on a muscle prior to its contraction were first described by Fenn in the 1920s. Later, Starling found that the isolated heart responded in the same manner, and this became widely known as Starling's law of the heart.

This basic characteristic of muscle response can be demonstrated by increasing the weight that a muscle must lift or by stretching the muscle in a progressive, stepwise manner. The method employed in this activity will stretch the muscle.

Effect of Stretch: Length-Tension Curve: Mount a metric ruler on the ring stand so that muscle length and stretch can be measured by sighting over the rod that holds the transducer (Figure 15.3). Use the tension adjuster to reduce the stretch on the muscle until a supramaximal voltage produces only a very small deflection (1–2 mm) when the muscle contracts. Record this as 0 mm of muscle stretched length. Record also the ruler measurement at this muscle length by sighting over the rod that holds the transducer. In **Input Amplifier**, make sure that the **AC Coupled** box is cleared. Adjust the range so that you can see a live trace in the window. Use a very slow speed (4/sec) for the rest of the activity.

Set up the **Stimulator** to stimulate the muscle with a single pulse of supramaximal voltage. Click **Start** and then **Stop** after the stimulus is delivered. Record the millimeters of stretch placed on the muscle, using the **Comments** window.

Stretch the muscle 1 mm and stimulate it as before. Continue in this manner, stretching the muscle in 1 mm increments, stimulating at each new length, and recording the millimeters of stretch associated with each contraction. Allow at least 10 sec of rest time after each stimulation before the muscle is stretched and stimulated again. Continue the experiment until no contraction occurs when the muscle is stimulated. *Keep the muscle moist with frog Ringer's during the entire experiment.* The record produced should resemble that shown in Figure 15.7.

From your record, determine the *passive* (resting) tension on the muscle in millimeters of pen deflection from the baseline position. Then determine the *active* (contractile) tension developed when the muscle was stimulated.

In the Laboratory Report, record the millimeters of passive and active tension developed at each muscle length. Then graph the data, with muscle stretched length on the abscissa (*x*-axis)

Copyright © 2015, 2011, 2008 Pearson Education, Inc.

and tension on the ordinate (y-axis). Plot all three tensions: active, passive, and total (active and passive) on the same graph, using different colors. At what stretched length does the muscle have the greatest active tension? At what stretched length is the active tension reduced to 0?

5. Neuromuscular Fatigue

Lift the sciatic nerve out from between the muscles and place it over the stimulating electrodes. Set **Stimulator** to deliver a supramaximal voltage at a tetanizing frequency (10–15/sec). Click **Start**. Continue stimulating until contraction fatigue is observed (contraction height is reduced by one-half). Click **Stop**.

Stimulate the nerve with a single shock of supramaximal voltage and record the strength (height) of the muscle contraction. Then place the electrodes directly on the muscle, stimulate it in the same manner, and record the twitch.

Allow the preparation to rest for 5 min and then repeat the nerve and muscle single-shock stimulations. Has complete recovery taken place?

6. Neuromuscular Blockade

Inject 0.1 ml of tubocurarine (3 mg/ml) into the belly of the muscle near the entrance of the nerve. An alternative method is to inject 1 ml of tubocurarine under the skin on the back of the frog so the drug can be absorbed by the lymphatic system into the blood.

In **Stimulator**, select the **Continuously, When Recordings Start**, and **PPM** options. Set the frequency to **6 PPM**, the stimulator amplitude to **Supramaximal** voltage, and pulse duration to **1 msec**. Click **Close**.

Using a very slow speed (4/sec), click **Start**. The nerve will be stimulated every 10 sec, using the same single supramaximal voltage used previously. Within 5–10 min, you should see a change in the strength of the muscle contraction.

When the contractions reach one-half the initial height, click **Stop**. Stimulate the muscle directly once again and record the contraction height. ◄

CLINICAL APPLICATION

Tetanus, Botulism, and Botox

Tetanus, a condition characterized by prolonged skeletal muscle contractions, is caused by tetanospasmin, a neurotoxin released by *Clostridium tetani*, an obligate anaerobic bacterium. Infections are associated with cuts or puncture wounds. Initial symptoms include lockjaw and facial spasms leading to a rigidity of pectoral and calf muscles. It acts by preventing the release of the inhibitory neurotransmitter GABA (gamma-aminobutyric acid). Even the slightest activity of motor neurons can result in dangerous hyperactivity of muscle fibers leading to prolonged and painful contractions. Botulism is another paralytic disease caused by the botulinum toxin produced by an infection of *Clostridium botulinum*. Infections result either from ingestion of contaminated food or from a wound. Severe infections can result in respiratory failure from paralysis of the breathing muscles. Botulism acts by degrading the SNAP-25 proteins required for the release of neurotransmitters at the axon terminal. The botulinum toxin (Botox) is best known for its cosmetic purposes but is also used in many more medical situations, including its use as a treatment for migraine, incontinence, strokes, Parkinson's disease, and vocal cord spasms, in each case reducing the contractions of muscles involved in the symptoms of the disease.

Human Myography

Simulation of Motor Points

The intact skeletal muscles of humans can be stimulated directly through the skin if a fairly strong stimulus is employed. Such stimulation is used to diagnose certain neuromuscular disorders and to prevent muscle atrophy during temporary paralysis in diseases such as poliomyelitis. Certain spots on the muscle are more sensitive to electrical stimulation than the rest of the muscle; these spots are called the *motor points*. The motor points usually lie over the point where the nerve enters the muscle; thus, the muscle contraction is produced through stimulation of the innervating nerve. Most motor points are located over the belly of the muscle.

► ACTIVITY 15.2

Stimulation of Motor Points

Materials

☐ 1 stimulator with a built-in isolation unit (Grass SD9)

☐ 1 plate electrode connected to the negative terminal of the stimulator

☐ 1 banana plug electrode connected to the positive terminal of the stimulator

☐ 1 tube electrode paste

1. Place electrode paste on an electrocardiogram (ECG) plate electrode and secure the electrode to the subject's upper arm with a rubber strap (Figure 15.11). Connect the plate electrode to the negative (−) output terminal of the electronic stimulator. This will serve as the reference electrode.

Copyright © 2015, 2011, 2008 Pearson Education, Inc.

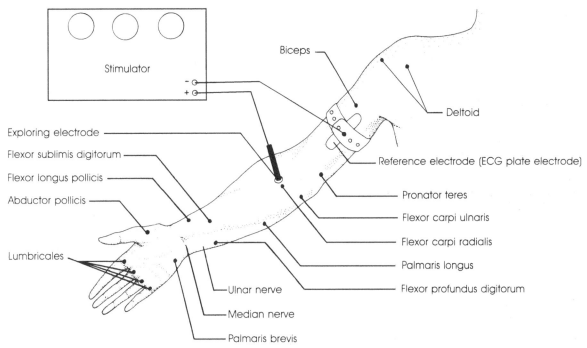

FIGURE 15.11   Stimulation of motor points.

2. Connect the exploring electrode to the positive (+) output terminal of the stimulator. A banana plug attached to a wire is used as the exploring electrode. Apply a small dab of electrode paste to the tip of the banana plug and reapply fresh paste as needed during the activity. **Note:** The best type of stimulator to use for motor point stimulation is one with three terminals: positive, negative, and chassis ground. On stimulators with only two terminals, the negative is common with the ground terminal. **If a two-terminal stimulator is used, the subject should not stimulate him- or herself because he or she might touch the stimulator, complete the electrical circuit, and receive a strong shock.**

3. Have the subject lay his or her forearm on a table with the flexor surface upward (palm up). Set the stimulator at 40–60 V, 1-msec duration, and repeat stimulations of one pulse per second. Using the exploring electrode, stimulate various points on the forearm to locate as many motor points as possible. Mark with ink each motor point that produces muscle twitches.

4. On the illustration of the arm provided in the Laboratory Report, indicate the location of the motor points you have stimulated and the threshold voltage of each motor point.

5. Locate the motor points for the flexor muscles of the fingers (for instance, flexor carpi ulnaris or radialis). Increase the voltage stepwise to show gradations of contraction up to a maximum. Then, using a maximal voltage, slowly increase the frequency of stimulation until tetanus is produced, but do not maintain tetanus longer than 2–3 sec. ◀

Human Myography

Muscle Recruitment and Fatigue

Muscle function in humans can be evaluated by examining contraction strengths using a hand dynamometer or by examining the depolarization of contracting muscles using surface electrodes connected through bioamplifiers to various data acquisition units. In the experiments that follow, you will initially correlate the electrical response of muscles under increasing degrees of generated force, using a hand dynamometer and surface electrodes. Because strength of contraction is correlated with the number of motor units being stimulated, we could expect to see increased electrical activity as contraction strength increases. Subsequently, you will examine the electrical response of some muscles while inducing fatigue.

Copyright © 2015, 2011, 2008 Pearson Education, Inc.

▶ ACTIVITY 15.3

Muscle Recruitment and Fatigue

15.3 VERNIER® VERSION

Materials (per group)

- ☐ 1 computer
- ☐ 1 Vernier® Logger Pro
- ☐ 1 hand dynamometer
- ☐ 1 set of EKG sensors
- ☐ 6 electrode tabs

Materials needed for this experiment will have been set up for you.

Students should work in pairs, with one playing the role of the experimental subject while the other is the investigator. It is important to shield the computer screen from the subject's view while an experiment is being performed.

Open the **18 EMG And Muscle Fatigue** file from the **Human Physiology With Vernier** folder.

1. Attach two electrode tabs to the ventral forearm of the dominant hand as shown in Figure 15.12. The third electrode tab should be placed on the upper arm. Attach the green and red leads to the tab on the forearm (order not important) and the black lead to the tab on the upper arm. These electrodes will measure the electrical activity of the muscles during contraction.
2. The hand dynamometer has to be calibrated by clicking the button on the toolbar on the screen. Hold the dynamometer along its sides, ensuring that no force is placed on the pads. Select the hand dynamometer by selecting the box and then click **OK**.
3. When the investigator clicks the **Collect** button, data is collected for 100 sec. During this time the subject will first perform three muscle contractions, using increasing strengths of contraction with a 15-sec break between contractions. The investigator will keep time and tell the subject when to contract and when to rest while controlling the computer.
4. The subject should look away from the screen and hold the selected arm at his or her side with the elbow resting on the table flexed at 90°. The dynamometer should be gripped in the palm.
5. The investigator will click **Collect** and instruct the subject (after a 5-sec delay) to perform the following:

    a. Weak contraction for 5 sec
    b. Rest for 15 sec
    c. Medium contraction for 5 sec
    d. Rest for 15 sec
    e. Maximal contraction for 5 sec

Your graph should look like that in Figure 15.13a. Save your file using a unique file name. You can print your results by clicking the graphs produced and choosing **File** and then **Print** from the menu options at the top of the screen.

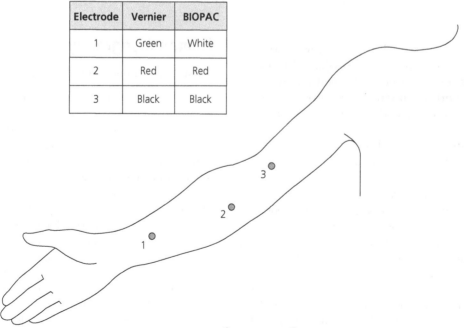

Electrode	Vernier	BIOPAC
1	Green	White
2	Red	Red
3	Black	Black

FIGURE 15.12 Electrode placements for Vernier® and BIOPAC® exercises.

Copyright © 2015, 2011, 2008 Pearson Education, Inc.

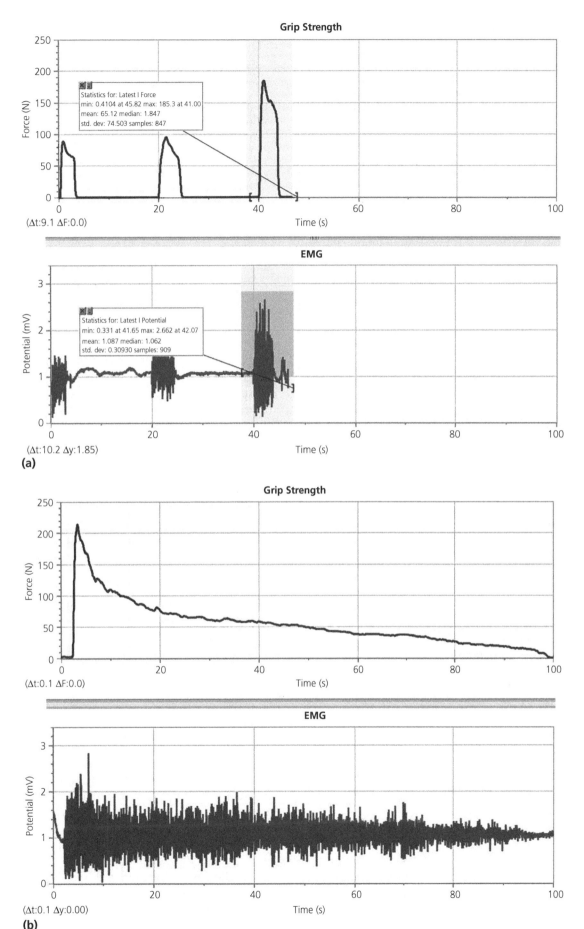

FIGURE 15.13 (a) Vernier® muscle recruitment screen capture. (b) Vernier® muscle fatigue screen capture.

Source: © Vernier Software & Technology. Used with permission.

Copyright © 2015, 2011, 2008 Pearson Education, Inc.

Data Analysis

You will use the Statistical tool to analyze your data and enter it into the table in the Laboratory Report. You will do this for the grip strength panel and the EMG panel.

1. Position your cursor at the beginning of the first contraction and highlight the entire contraction by clicking and dragging. Click the **Statistics** button and record the mean force for that contraction, rounding up to the nearest 0.1N. Enter this information in the table in the Laboratory Report.

   In the **EMG** panel, repeat the process and record the maximum mV and minimum mV associated with each contraction. Calculate the difference between these values and enter this information into the column headed Δ **mV**.

   After resting for 2 min, click **Collect** once again, but this time start with a maximal strength of contraction and sustain your grip for as long as possible. Relax your grip when completely fatigued. Your graph should look like Figure 15.13b. Save your file using a unique file name. Analyze your data as you did in the previous step.

   Enter your data and answer the questions in the Laboratory Report. You can print your results by clicking the graphs produced and choosing **File** and then **Print** from the menu options at the top of the screen.

   You can repeat both experiments using your nondominant arm.

15.3 BIOPAC® VERSION

Materials (per group)

☐ 1 computer

☐ 1 BIOPAC® Student Lab software PC 3.7.0 or Mac 3.0.7 or later

☐ 1 BIOPAC® data acquisition unit

☐ 1 hand dynamometer

☐ 1 electrode lead set

☐ 6 disposable electrode tabs

Materials needed for this experiment will have been set up for you.

Students should work in pairs, with one playing the role of the experimental subject while the other is the investigator. It is important to shield the computer screen from the subject's view while an experiment is being performed.

Start the BIOPAC® Student Lab Program and choose lesson **L02-EMG-2**.

1. Attach two electrode tabs to the ventral forearm of the dominant hand as shown in Figure 15.12. The third electrode tab should be placed on the upper arm. Attach the white (−) lead to the forearm tab nearest the elbow, the red (+) lead to the tab on the forearm nearest the wrist, and the black lead (ground) to the tab on the upper arm. These electrodes will measure the electrical activity of the muscles during contraction.

2. Click **Calibrate** and follow the onscreen instructions to calibrate the dynamometer. Click **OK** when ready. When the calibration recording begins, wait 2 sec before clenching the hand dynamometer as hard as possible and then releasing. The calibration procedure will last 8 sec and then stop. You are now ready to proceed with the experiment.

3. When the investigator clicks the **Continue/Record** button, the experiment will start. During this time, the subject will first perform three muscle contractions using increasing strengths of contraction with a 15-sec break between contractions. The investigator will keep time and tell the subject when to contract and when to rest while controlling the computer.

4. The subject should now look away from the screen and hold the selected arm with the elbow resting on the table, flexed at 90°. The dynamometer should be gripped in the palm.

5. The investigator will click **Continue/Record** and instruct the subject (after a 5-sec delay) to perform the following (in order):

   a. Weak contraction for 5 sec
   b. Rest for 15 sec
   c. Medium contraction for 5 sec
   d. Rest for 15 sec
   e. Maximal contraction for 5 sec

6. Click **Suspend**.

Your data should look like that in Figure 15.14a.

After resting for 2 min, click **Continue/Resume** and, after a 5-sec delay, start with a maximal strength of contraction and sustain your grip for as long as possible. Relax your grip when completely fatigued. Click **Suspend** to halt the recording. Your data should resemble that in Figure 15.14b.

Click **Continue/Resume** and repeat both experiments using your nondominant arm.

Click **Stop** and then click **Done** when all experiments are complete.

Copyright © 2015, 2011, 2008 Pearson Education, Inc.

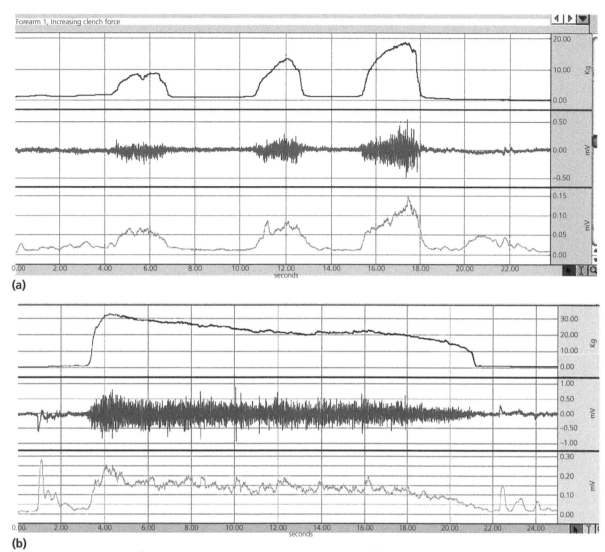

(a)

(b)

FIGURE 15.14   (a) BIOPAC® muscle recruitment screen capture. (b) BIOPAC® muscle fatigue screen capture.

Data Analysis

Enter **Review Saved Data** from the **Lessons** menu.

Depending on the version of the software being used, three (or two) panels will be visible, labeled **Force**, **EMG**, and (hidden) **Integrated EMG**. We will be examining the data in the first two panels. The measurement boxes are above the marker region in the data window. Each measurement has three sections: Channel Number, Measurement Type, and Result. The first two are pull-down menus that are activated when clicked. For each channel, choose **Max** in the measurement type. You will use the I-beam tool at the bottom of the data panel to select a portion of the graph.

1. In the segment labeled **Forearm 1, Increasing Clench Force,** select sequentially (by clicking and dragging) each complete clench force, using the I-beam tool. Read the **Max** value for Channel 1 (Force) and Channel 3 (EMG) and enter the information in the table in your Laboratory Report. The minimum value for each of these should be zero.

2. Repeat this for the segments labeled **Forearm 1, Continued Clench At Maximum Force**, **Forearm 2, Increasing Clench Force**, and **Forearm 2, Continued Clench At Maximum Force**. Enter this information in the table in your Laboratory Report and answer the questions that follow. ◄

Copyright © 2015, 2011, 2008 Pearson Education, Inc.

Nerve–Muscle Activity

Name _____

Date _____ Section _____

Score/Grade_____

Isolated Muscle Responses

1. Muscle and nerve irritability

Threshold voltage with nerve stimulation _____

Threshold voltage with direct muscle stimulation _____

Why is there a difference in these threshold voltages?

_____

_____

What is meant by the independent irritability of muscle?

_____

_____

2. Effects of stimulus strength

Place your record in the following space.

How is the response obtained related to the all-or-none law of muscle fiber contraction?

_____

_____

3. Effects of stimulus frequency

Place your record in the following space.

Explain the mechanism responsible for summation of contractions and the increase in height of contraction when the stimulus frequency is increased.

_____

_____

Copyright © 2015, 2011, 2008 Pearson Education, Inc.

What is tetanus? Why is it produced?

_____

_____

4. Effects of stretch or load on muscle contractility

    a. Effect of stretch: length-tension curve (Physiograph™ system)

       Place your record in the following space.

Muscle Stretch Length (mm)	Passive Tension (mm)	Active Tension (mm)	Total Tensions (mm)

Copyright © 2015, 2011, 2008 Pearson Education, Inc.

Tension (mm pen deflection)

Muscle stretch length (mm)

How is the contractile tension developed with increasing stretch related to the sarcomere structure of the muscle fiber?

_____

_____

Of what practical importance is this response of muscle to stretch in the *in vivo* situation?

_____

_____

b. Effect of stretch: length-tension curve (PowerLab® system)

Place your record in the following space.

Copyright © 2015, 2011, 2008 Pearson Education, Inc.

(vertical axis label) Tension (mm trace deflection)

Muscle stretch length (mm)

How is the contractile tension developed with increasing stretch related to the sarcomere structure of the muscle fiber?

_____

_____

Of what practical importance is this response of muscle to stretch in the *in vivo* situation?

_____

_____

5. Neuromuscular fatigue

Record the height of the muscle contractions in millimeters of pen deflection.

Immediately after fatigue		**After 5 min of rest**	
Nerve stimulated	Muscle stimulated	Nerve stimulated	Muscle stimulated
_____	_____	_____	_____

Explain the difference in contraction height obtained with nerve and muscle stimulation immediately after fatigue. Where does fatigue occur first? Why?

_____

_____

Which has benefitted most from the rest period—the nerve or the muscle? Explain the processes that enable this recovery.

_____

_____

Copyright © 2015, 2011, 2008 Pearson Education, Inc.

6. Neuromuscular blockade

Record the height of the muscle contractions before and after tubocurarine injection.

Before tubocurarine injection		**After tubocurarine injection**	
Nerve stimulated	Muscle stimulated	Nerve stimulated	Muscle stimulated
_____	_____	_____	_____

How do you explain the differences in contraction height? What is the mechanism of action of tubocurarine?

_____

_____

Of what clinical use are drugs such as tubocurarine?

_____

_____

Stimulation of Motor Points

1. On the following diagram, indicate the location of the motor points you have stimulated and the threshold voltage at each point.

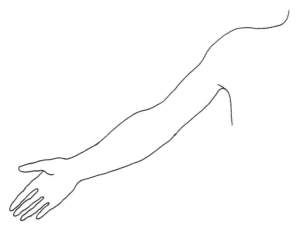

2. At what frequency does tetanus occur when the finger flexor muscles are stimulated?

_____

3. How does this frequency compare with the frequency needed to tetanize the frog gastrocnemius muscle you worked with previously?

_____

4. What is a motor unit?

_____

_____

5. How does the anatomical composition of motor units controlling different muscle groups vary? What is the purpose of this variation?

_____

_____

Copyright © 2015, 2011, 2008 Pearson Education, Inc.

6. What is the effect of curare on muscle contraction when the nerve is stimulated? What is curare's mechanism of action? How could a large dose of curare cause death?

_____

_____

7. Describe and explain what happens to muscle contraction when the nerve is stimulated or the muscle is stimulated directly while either is in a calcium-free bath.

_____

_____

Human Myography

Dominant Arm

Time Interval	Grip Strength (N or Kg)	EMG Data		
	Mean Grip Strength	Max mV	Min mV	ΔmV
5–10 sec (weak contraction)				
25–30 sec (medium contraction)				
45–50 sec (maximum contraction)				
Sustained Contraction				

Nondominant Arm

Time Interval	Grip Strength (N or Kg)	EMG Data		
	Mean Grip Strength	Max mV	Min mV	ΔmV
5–10 sec (weak contraction)				
25–30 sec (medium contraction)				
45–50 sec (maximum contraction)				
Sustained Contraction				

8. Of the first three contractions, which contraction generated the smallest maximum voltage? Which contraction generated the greatest maximum voltage?

_____

_____

Copyright © 2015, 2011, 2008 Pearson Education, Inc.

9. Using your knowledge of muscle physiology, explain the difference between the first three contractions on the basis of motor unit recruitment.

_____

_____

10. Using your graphical record of the sustained contraction, describe what happened to your grip strength as you progressed through the entire trial.

_____

_____

11. Using your graphical record of the EMG associated with your sustained contraction, describe the changes you observed as you progressed through the entire trial.

_____

_____

12. Explain your observations, based on your knowledge of recruitment and muscle fatigue.

_____

_____

_____

_____

13. Compare the response of your dominant versus your nondominant arm in terms of contraction strength and millivolts generated during maximal contraction. Explain the difference, using motor unit size and recruitment differences between both arms.

_____

_____

_____

APPLY WHAT YOU KNOW

1. Based on your knowledge of muscle physiology and the nature of motor units, how would the structure of muscles in your thigh differ from muscles controlling the movements of your eyeball (the extraocular muscle)?

_____

_____

2. Imagine holding a heavy object stationary in front of you with your arm fully extended at the elbow. Describe, at the level of a sarcomere in which actin/myosin interactions are taking place, how you maintain this position. Because no movement is taking place, are you using ATP? If yes, explain your answer.

_____

_____

_____

_____

Copyright © 2015, 2011, 2008 Pearson Education, Inc.

Cardiac Function

CHAPTER 16 INCLUDES:

PowerLab 1 PowerLab® Activity

PhysioEx 9.1 For more exercises on Cardiac Function, visit PhysioEx™ (www.physioex.com) and choose Exercise 6: Cardiovascular Physiology.

⚠ CAUTION!

Because these experiments involve the use of living vertebrates, it is important that you obtain the appropriate permissions from your Institutional Animal Care and Use Committee (IACUC). While the instructions given are suitable to obtain the desired result, your IACUC may require a modification of procedures to meet your institution's needs.

OBJECTIVES

After completing this exercise, you should be able to

1. Explain the mechanisms regulating contractility of cardiac musculature.
2. Differentiate between mammalian and amphibian hearts.
3. Examine the normal contraction pattern of the amphibian heart.
4. Demonstrate the effects of sympathetic and parasympathetic stimulation of the heart.
5. Mimic the effects of heart blocks on the frog heart.

Characteristics of Heart Contractility

The heart's primary function is simply to act as a pump that provides pressure to move blood to its ultimate destination—the tissues. The control of cardiac contractility is complex and represents a balance of **intrinsic** (within the heart) and **extrinsic** (outside the heart) factors. In the following activity, you will examine some of these intrinsic and extrinsic factors that make the heart such a unique and versatile pump.

Although some similarities exist in the structure and function of skeletal and cardiac muscle cells, a number of very significant differences are also apparent. Cardiac muscle cells are uninucleate and branched, whereas skeletal muscle cells are multinucleate and unbranched. Additionally, skeletal muscles cells require motor innervations to contract, whereas cardiac muscle cells do not. In fact, cardiac muscles are capable of automaticity, the ability to initiate their own rhythmic contractions. The coordination of muscular contractions in the atrial and ventricular muscle syncytia emerge from the presence of autorhythmic myocardial cells generating action potentials. These action potentials spread through the specialized non-contractile myocardial fibers that make up the cardiac conduction system. Myocardial autorhythmic cells generate action potentials due to their unstable membrane potential, which, at −60 mV, starts depolarizing due to the opening of I_f channels, which are permeable to both K^+ and Na^+. Because Na^+ movement in exceeds K^+ movement out, the cells depolarize, opening slow Ca^{2+} channels, which then raises the membrane potential to threshold. This produces the generator potential. At this time, fast Ca^{2+} channels open, creating a steep depolarization of the autorhythmic cell. This is the phase of rapid depolarization. When these Ca^{2+} channels close, slow K^+ channels open and repolarize the cells to −60 mV, restarting the cycle. Figure 16.1 illustrates these changes.

In the mammalian heart, the pacemaker is the **sinoatrial (SA) node**, a group of specialized cells near the junction of the vena cava and the right atrium. In the frog or turtle heart, the pacemaker is the **sinus venosus**, an enlarged region between the vena cava and the right atrium. (The mammalian SA node is believed to be an evolutionary remnant of the sinus venosus.) Each region of the heart has its own intrinsic rate of beating: for example, SA node, 100 beats/min;

Copyright © 2015, 2011, 2008 Pearson Education, Inc.

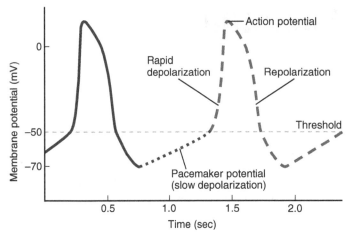

FIGURE 16.1 Electrical activity in the SA node pacemaker cells.

atrium, 60 beats/min; ventricle, 35 beats/min. Only when the faster pacemaker region is blocked is it possible to observe the intrinsic rate of the slower regions. Autonomic innervation of the SA node can change the heart rate by either increasing or decreasing the permeability of the autorhythmic cells to Na^+ and Ca^{2+} or K^+. Sympathetic stimulation, using adrenergic neurotransmitters, speeds up the heart rate by increasing the ionic flow through the I_f and the Ca^{2+} channels. The parasympathetic secretion of acetylcholine will hyperpolarize the SA node cells by increasing K^+ permeability and decreasing Ca^{2+} permeability, reducing the heart rate. The effects of sympathetic and parasympathetic stimulation are illustrated in Figure 16.2.

Action potentials generated in the pacemaker cells are conducted through the cardiac conduction system to the myocardial contractile cells. Here the action potential spreads through gap junctions connecting adjacent cells. Although action potentials in different parts of the heart vary, the sequence of events is rather similar in that the initial depolarization is brought about by opening voltage-gated sodium channels and the movement of Na^+ into the cells. This is a very rapid state of depolarization and raises the membrane potential to between $+30$ and $+40$ mV. When the sodium gates inactivate, repolarization begins due to the opening of potassium outflux, but this is quickly balanced by calcium channels opening, allowing the influx of Ca^{2+}. This extends the period of repolarization into a plateau phase, effectively keeping the cell in a depolarized state. When the Ca^{2+} channels close, the opening of additional K^+ channels repolarizes the contractile cells. This sequence of events is illustrated in Figure 16.3. Other intrinsic characteristics of cardiac contraction are shown in the following activity.

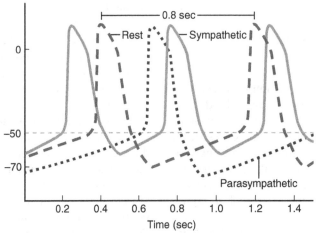

FIGURE 16.2 The effects of sympathetic and parasympathetic stimulation on the SA nodal cell potentials.

Copyright © 2015, 2011, 2008 Pearson Education, Inc.

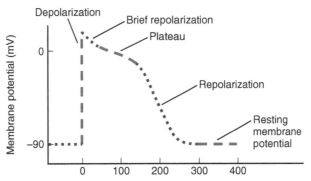

FIGURE 16.3   The cardiac contractile cell action potential.

CLINICAL APPLICATION

When Blood Pressure Becomes a Risk Factor

Hypertension is a medical condition in which an individual experiences chronic high blood pressure. Whereas normal blood pressure is less than 120/80 mm of Hg (systolic/diastolic), an individual with hypertension has blood pressure that consistently exceeds 140 mm systolic pressure or 90 mm diastolic pressure. Hypertension is a risk factor for a number of other medical afflictions, including heart failure, coronary artery disease, renal disease, and peripheral artery disease, among others. This makes blood pressure control an important preventive intervention. A number of strategies can be used to reduce blood pressure, some as simple as lifestyle changes involving diet (reducing salt intake), exercise, weight loss, and quitting smoking tobacco and drinking alcohol. Blood pressure can also be treated using medications that either reduce the contraction strength of the heart while dilating blood vessels (beta blockers and ACE inhibitors) or reduce blood volume by increasing urine production (diuretics).

Anatomy of Amphibian or Reptilian Heart

In the following activity, you will use a frog or turtle heart because it functions well at room temperature and will continue to beat even when excised from the body. Mammalian hearts have the same contractile characteristics but must be supplied with a constant flow of warm, oxygenated blood to maintain their contractility.

The frog and turtle hearts differ from the mammalian heart anatomically in that they are three-chambered rather than four-chambered. The pacemaker in the amphibian heart is the sinus venosus, a thin-walled sac that receives blood from the anterior and posterior caval veins and empties blood into the right atrium. The single ventricle receives blood from both atria and pumps blood out through the large artery called the **truncus arteriosus** (Figure 16.4). In contrast, the mammalian ventricle has separate left and right chambers, which prevent mixing of the venous and arterial blood (Figure 16.5).

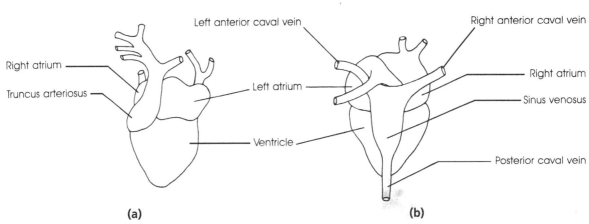

FIGURE 16.4   Frog heart. (a) Ventral view. (b) Dorsal view.

Copyright © 2015, 2011, 2008 Pearson Education, Inc.

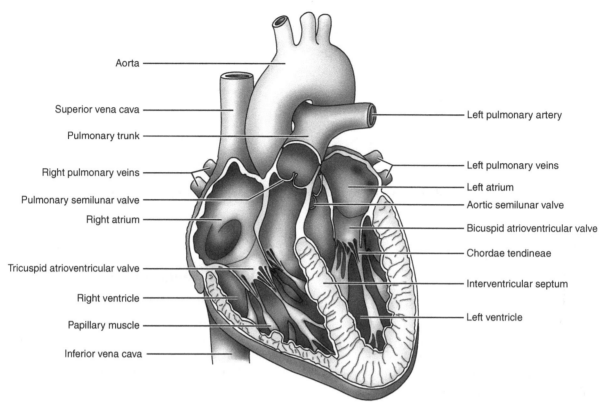

Aorta

Superior vena cava

Pulmonary trunk

Right pulmonary veins

Pulmonary semilunar valve

Right atrium

Tricuspid atrioventricular valve

Right ventricle

Papillary muscle

Inferior vena cava

Left pulmonary artery

Left pulmonary veins

Left atrium

Aortic semilunar valve

Bicuspid atrioventricular valve

Chordae tendineae

Interventricular septum

Left ventricle

FIGURE 16.5   Mammalian heart.

Frog Heart Dissection Preparation

Double pith a frog and fasten it to a frog board, ventral side up. Use scissors to make a longitudinal incision through the skin and body wall of the thoracic region to expose the heart. Note the pericardial sac surrounding the heart. Hold the pericardium with forceps and, using scissors, carefully cut away the sac from the heart. From this point on, make sure that the heart is periodically moistened with frog Ringer's solution.

Using forceps, gently lift the apex of the heart upward. Insert a bent insect pin or small fishhook through the tip of the ventricle, being careful not to damage the ventricle. Tie a thin thread to the hook and connect the ventricle to the transducer just as you connected the gastrocnemius muscle to the transducer in Chapter 15. If you use the Physiograph™ transducer, attach the thread to the transducer hook and adjust the tension on the ventricle until the recording pen is raised slightly above the baseline (Figure 16.6).

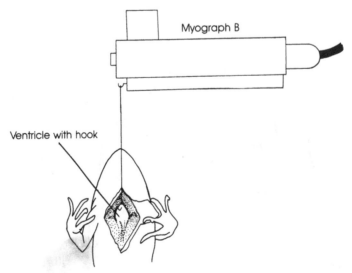

Myograph B

Ventricle with hook

FIGURE 16.6   Frog heart preparation: Physiograph™ setup.

Copyright © 2015, 2011, 2008 Pearson Education, Inc.

Turtle Heart Dissection Preparation

Using a heavy forceps, grab the turtle's upper beak and pull the head out. Slip a ligature of heavy cord over the head and pull it tightly around the neck. Destroy the brain by pithing it with a heavy probe. It is also possible to anesthetize the turtle by injecting 2 ml of 3% pentobarbital solution (Nembutal) into the peritoneal cavity (insert the needle between the hind limbs and the body).

Cut a circular opening in the plastron (ventral shield) at the level of the heart, using a drill press fitted with a hole saw. If a motorized saw is not available, cut away the entire plastron using a hand saw. This latter method is not as desirable because of the chance of hitting major blood vessels and causing excessive hemorrhage.

Attach ligatures to each of the turtle's legs and tie the head and legs to a turtle board. Cut away the pericardial sac around the heart. Tie a thread around the frenulum cordis near the apex of the ventricle, cut the frenulum peripherally, and attach the thread to a heart lever or transducer. Keep the heart moist with frog Ringer's solution.

Physiology of Amphibian or Reptilian Heart

▶ ACTIVITY 16.1

Heart Physiology Using Physiograph™

Materials

☐ 1 Physiograph™

☐ 1 electronic myograph transducer

☐ 1 set stimulating electrodes

☐ 1 frog

☐ 2 glass probes

☐ 1 dissecting needle

☐ 1 6-inch metric ruler

☐ 1 10-cm piece of cotton string

☐ 1 250-ml squeeze bottle of frog Ringer's solution at room temperature

☐ 1 250-ml squeeze bottle of frog Ringer's solution at 40°C

☐ Ice bath to cool Ringer's solution to 5°C

☐ Dissection equipment (forceps, scissors, bone cutters)

☐ 5 ml acetylcholine (0.1 mg/ml)

☐ 5 ml epinephrine (1 mg/ml)

☐ 5 ml pilocarpine (0.2 mg/ml)

☐ 5 ml nicotine (1 mg/ml)

☐ 5 ml atropine (1 mg/ml)

1. Normal Heartbeat

Obtain a recording of the normal cardiac rhythm, using a medium to fast paper speed to distinguish the atrial and ventricular contractions. Run a 1-sec time line while recording so that the duration of systole and diastole of the ventricle can be determined. Your record should resemble that shown in Figure 16.7. Attach a labeled portion of your record to the Laboratory Report.

2. Refractory Period of Heart

Position the transducer to eliminate as much as possible the atrial contraction in the recording. Arrange for electrical stimulation of the ventricle by clamping the stimulating electrode so that the points touch the ventricle gently and constantly during the contraction cycle. An alternative method is to connect the stimulator to the heart by fine copper wires. Wrap one wire around the base of the ventricle and the other around the apex of the ventricle, thus providing a closed circuit through the length of the ventricle.

Record the ventricular contractions, using a medium paper speed. Using a single 20 V stimulus of 1-msec duration, stimulate the ventricle at different times in the cardiac cycle as shown in Figure 16.8. Begin at time six and work backward through the cycle.

What is the result of stimulating during the systolic phase of the cycle? During the diastolic phase? Can a second contraction be elicited before the normal rhythmic contraction occurs? Look for the appearance of an extra systole followed by a compensatory pause as shown in Figure 16.9.

It is often tricky to obtain an extra systole using single stimuli because it is difficult to catch the

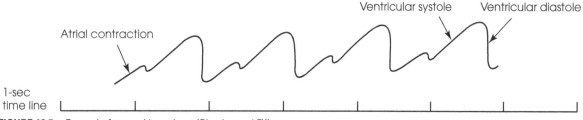

FIGURE 16.7   Record of normal heartbeat (Physiograph™).

Copyright © 2015, 2011, 2008 Pearson Education, Inc.

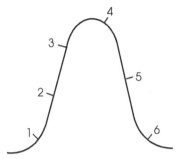

FIGURE 16.8 Single ventricular myogram. The numbers 1 through 6 denote times of stimulus during a heart cycle.

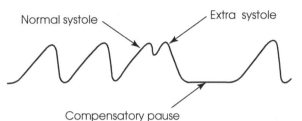

FIGURE 16.9 Extra systole.

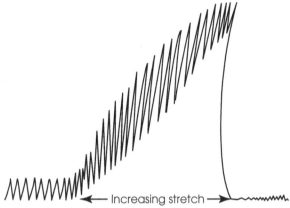

FIGURE 16.10 Contractions of a stretched ventricle.

ventricle immediately after its refractory period. If you have trouble, try using repeated stimulation so that one of the stimuli can catch the ventricle at the proper time to produce an extra systole.

3. Tetanization of Heart
Record the contraction of the heart while it is being stimulated with a tetanizing frequency (10–15 stimuli per second) and 20 V. Remove the stimulating electrodes after this activity.

Record your observations and answer the questions in the Laboratory Report.

4. Starling's Law of the Heart
In the early 1900s, Ernest Starling's investigations revealed that the energy of contraction is proportional to the initial length of the cardiac muscle fiber. This statement became known as Starling's law of the heart, a major concept in cardiovascular physiology. In the intact animal, the length of the cardiac muscle fiber is increased by an increase in diastolic filling of the heart. In this activity, you will increase the fiber length by simply stretching the ventricle.

Adjust the transducer sensitivity so that the height of the ventricular myogram is 2 cm. Position the recording pen so that contractions are recorded near the bottom of the pen excursion range.

Using a medium paper speed, begin stretching the ventricle incrementally by turning the tension adjuster. Continue stretching the ventricle until no additional change is noted. The record obtained should resemble that shown in Figure 16.10.

Record your observations and answer the questions in the Laboratory Report.

5. Temperature Effects
Record the heart contractions at room temperature. Then drop warm (40°C) frog Ringer's solution on the heart until significant changes are seen in rate and contractility. Record contractions at this time. Finally, drop cold (5°C) Ringer's on the heart

and record when changes are observed. Then rinse the heart with room temperature Ringer's to return the beat to normal before continuing the activity. Can you see why the ectothermic animal becomes so sluggish when the temperature drops? Determine the heart rate at each temperature.

Record your observations and answer the questions in the Laboratory Report.

6. Drug Effects
Using a syringe or medicine dropper, apply the following drugs to the heart until you see significant changes in rate, contractility, or tone (changes in baseline). Best results are seen when the drug is dropped on the sinus venosus region of the heart. Be very cautious in applying these drugs because they are very potent and can stop the heart completely if an overdose is given. Acetylcholine and nicotine are especially potent. If the heart does stop, apply epinephrine to restore the beat. After the effect is recorded, rinse the heart with frog Ringer's solution and allow the heart to return to normal before the next drug is applied.

a. Acetylcholine: 0.1 mg/cc (1:10,000). This is the normal transmitter released by the vagus (parasympathetic) nerve innervating the heart.

b. Epinephrine: 1 mg/cc (1:1000). This is an analog of norepinephrine, released by the postganglionic sympathetic nerves innervating the heart.

Copyright © 2015, 2011, 2008 Pearson Education, Inc.

c. Pilocarpine: 0.2 mg/cc (1:5000). This is an alkaloid drug from the leaf of *Pilocarpus jaborandi*. It acts like acetylcholine by directly stimulating muscarinic receptors in effector organs.

d. Nicotine: 1 mg/cc (1:1000). This is an alkaloid from the tobacco plant. It acts on the autonomic ganglia, combining with the acetylcholine receptor on the postganglionic neuron. In small doses, nicotine stimulates synaptic transmission in the ganglia, whereas in larger doses, it depresses synaptic activity. Because the terminal ganglion (parasympathetic division) lies in the heart muscle, this ganglion can be affected by dropping nicotine directly on the heart. Could the sympathetic ganglia also be affected by this application?

e. Atropine: 1 mg/cc (1:1000). This is an alkaloid from *Atropa belladonna* that blocks the receptors for acetylcholine. After applying atropine and observing its effect, drop some acetylcholine on the heart and compare the heart's response with that seen previously when you applied acetylcholine.

Record your observations and answer the questions in the Laboratory Report.

7. Heart Block

Using a heavy thread, tie a loose single-loop ligature around the heart at the junction of the atria and ventricle. Take two short pieces of thread and run each one through the ligature on opposite sides of the loop (Figure 16.11). A Gaskell clamp may also be used to apply pressure on the atrioventricular junction.

Tighten the ligature slowly and observe the beating of the atria and ventricle for changes in rhythm. Do not tighten the ligature too much or you will cut the heart and damage it irreversibly. Can you see any of the following types of heart block?

First-degree block The interval between atrial and ventricular contraction is prolonged.

Second-degree block Some impulses fail to reach the ventricle so that the ratio of atrial to ventricular

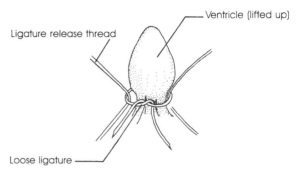

FIGURE 16.11   Heart ligature.

beats is altered. You might be able to see 2:1, 3:1, 5:1, and 8:1 types of heart block (Figure 16.12).

Third-degree block Impulses fail to pass through the atrioventricular (AV) node and bundle of His, and the ventricle might start its own independent rhythm of beating, or you might see only the atria contracting.

After producing a complete (third-degree) heart block, release the ligature by pulling on the release threads and observe whether a normal AV beat is restored.

Record your observations and answer the questions in the Laboratory Report.

8. All-or-None Law of the Heart

Increase the pressure on the AV ligature until the ventricle is completely quiet (no impulses are reaching the ventricle). Using electrodes, stimulate the ventricle to determine the threshold stimulus. Record the contraction height. Using a slow paper speed, stimulate the ventricle with increasing voltages (single stimuli), recording the contraction height at each voltage.

16.1 POWERLAB® VERSION

Materials (per group)

☐ 1 computer

☐ 1 PowerLab® data acquisition unit

☐ 1 force transducer

☐ 1 set stimulating electrodes

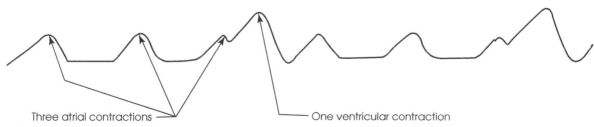

FIGURE 16.12   A 3:1 second-degree heart block.

Copyright © 2015, 2011, 2008 Pearson Education, Inc.

☐ 1 frog

☐ 2 glass probes

☐ 1 dissecting needle

☐ 1 6-inch metric ruler

☐ 1 10-cm piece of cotton string

☐ 1 250-ml squeeze bottle of frog Ringer's solution at room temperature

☐ 1 250-ml squeeze bottle of frog Ringer's solution at 40°C

☐ Ice bath to cool Ringer's solution to 5°C

☐ Dissection equipment (forceps, scissors, bone cutters)

☐ 5 ml acetylcholine (0.1 mg/ml)

☐ 5 ml epinephrine (1 mg/ml)

☐ 5 ml pilocarpine (0.2 mg/ml)

☐ 5 ml nicotine (1 mg/ml)

☐ 5 ml atropine (1 mg/ml)

Materials needed for this exercise will have been set up for you. You now set up the Chart software, as you did in Chapter 15, to investigate frog or turtle heart contraction.

1. Normal Heartbeat

Click **Start** to obtain a recording of the normal cardiac rhythm, using a medium to fast speed (200/sec) to distinguish the atrial and ventricular

contractions. Use the cursor and the record obtained to determine the duration of systole and diastole of the ventricle. Your record should resemble that shown in Figure 16.13. Attach a labeled portion of your record to the Laboratory Report.

2. Refractory Period of Heart

Position the transducer to eliminate as much as possible the atrial contraction in the recording. Arrange for electrical stimulation of the ventricle by clamping the stimulating electrode so that the points touch the ventricle gently and constantly during the contraction cycle. An alternative method is to connect the stimulator to the heart by fine copper wires. Wrap one wire around the base of the ventricle and the other around the apex of the ventricle, thus providing a closed circuit through the length of the ventricle.

Click **Start** to record the ventricular contractions, using a medium speed (100/sec). Set **Stimulator** to deliver a single stimulus of 10 V and 1-msec duration, and select **Start Manually** to allow you to determine when the stimulus is delivered. Close the stimulator window and click **Start**. Choose **Stimulator Panel** in **Setup**. Use the **Stimulate** button to stimulate the ventricle at different times in the cardiac cycle as shown in Figure 16.7. Begin at time six and work backward through the cycle.

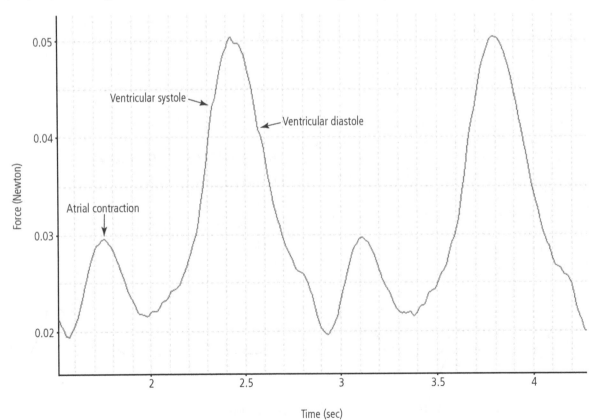

FIGURE 16.13   Record of normal heartbeat (PowerLab®).

Copyright © 2015, 2011, 2008 Pearson Education, Inc.

What is the result of stimulating during the systolic phase of the cycle? During the diastolic phase? Can a second contraction be elicited before the normal rhythmic contraction occurs? Look for the appearance of an extra systole followed by a compensatory pause, as shown in Figure 16.9.

It is often tricky to obtain an extra systole using single stimuli because it is difficult to catch the ventricle immediately after its refractory period. If you have trouble, try using repeated stimulation so that one of the stimuli can catch the ventricle at the proper time to produce an extra systole.

3. Tetanization of Heart

Record the contraction of the heart while it is being stimulated with a tetanizing frequency (10–15 stimuli per second) and 20 V. Print the graph obtained. Remove the stimulating electrodes after this activity.

Record your observations and answer the questions in the Laboratory Report.

4. Starling's Law of the Heart

In the early 1900s, Ernest Starling's investigations revealed that the energy of contraction is proportional to the initial length of the cardiac muscle fiber. This statement became known as Starling's law of the heart, a major concept in cardiovascular physiology. In the intact animal, the length of the cardiac muscle fiber is increased by an increase in diastolic filling of the heart. In this activity, you will increase the fiber length by simply stretching the ventricle.

In the **Input Amplifier** (see Chapter 15, Figure 15.8), make sure that the **AC Coupled** box is cleared. Adjust the range so that you can see a live trace in the window.

Using a medium speed (100/sec), begin stretching the ventricle incrementally by turning the tension adjuster.

Click **Start** and then **Stop** after a single atrial and ventricular contraction has been recorded. Record the millimeters of stretch placed on the heart using the comments window.

Continue stretching the ventricle and recording the response of the heart until no additional change in contraction is noted. You can add comments indicating the change in millimeters of stretch by typing in the comments box and pressing **Enter** on your computer keyboard while recording data. The record obtained should resemble that shown in Figure 16.10.

Record your observations and answer the questions in the Laboratory Report.

5. Temperature Effects

Record the heart contractions at room temperature. Then drop warm (40°C) frog Ringer's solution on the heart until significant changes are seen in rate and contractility. Click **Start** to record contractions at this time. Finally, drop cold (5°C) Ringer's on the heart and record when changes are observed. Then rinse the heart with room temperature Ringer's to return the beat to normal before continuing the activity. Can you see why the ectothermic animal becomes so sluggish when the temperature drops? Determine the heart rate at each temperature. Click **Stop** after these tests.

Record your observations and answer the questions in the Laboratory Report.

6. Drug Effects

Using a syringe or medicine dropper, apply the following drugs to the heart until you see significant changes in rate, contractility, or tone (changes in baseline). Best results are seen when the drug is dropped on the sinus venosus region of the heart. Be very cautious in applying these drugs because they are very potent and can stop the heart completely if an overdose is given. Acetylcholine and nicotine are especially potent. If the heart does stop, apply epinephrine to restore the beat. After the response is recorded, rinse the heart with frog Ringer's solution and allow the heart to return to normal before the next drug is applied. Use the comments window to indicate the drug used.

a. Acetylcholine: 0.1 mg/cc (1:10,000). This is the normal transmitter released by the vagus (parasympathetic) nerve innervating the heart.
b. Epinephrine: 1 mg/cc (1:1000). This is an analog of norepinephrine, released by the postganglionic sympathetic nerves innervating the heart.
c. Pilocarpine: 0.2 mg/cc (1:5000). This is an alkaloid drug from the leaf of *Pilocarpus jaborandi*. It acts like acetylcholine by directly stimulating muscarinic receptors in effector organs.
d. Nicotine: 1 mg/cc (1:1000). This is an alkaloid from the tobacco plant. It acts on the autonomic ganglia, combining with the acetylcholine receptor on the postganglionic neuron. In small doses, nicotine stimulates synaptic transmission in the ganglia, whereas in larger doses, it depresses synaptic activity. Because the terminal ganglion (parasympathetic division) lies in the heart muscle, this ganglion can be affected by dropping nicotine directly on the heart. Could the sympathetic ganglia also be affected by this application?

Copyright © 2015, 2011, 2008 Pearson Education, Inc.

e. Atropine: 1 mg/cc (1:1000). This is an alkaloid from *Atropa belladonna* that blocks the receptors for acetylcholine. After applying atropine and observing its effect, drop some acetylcholine on the heart and compare the heart's response with that seen previously when you applied acetylcholine.

Record your observations and answer the questions in the Laboratory Report.

7. Heart Block

Using a heavy thread, tie a loose single-loop ligature around the heart at the junction of the atria and ventricle. Take two short pieces of thread and run each one through the ligature on opposite sides of the loop (Figure 16.11). A Gaskell clamp may also be used to apply pressure on the atrioventricular junction.

Tighten the ligature slowly and observe the beating of the atria and ventricle for changes in rhythm. Do not tighten the ligature too much or you will cut the heart and damage it irreversibly. Click **Start** to record the contractions. Can you see any of the following types of heart block?

First-degree block The interval between atrial and ventricular contraction is prolonged.

Second-degree block Some impulses fail to reach the ventricle so that the ratio of atrial to ventricular beats is altered. You might be able to see 2:1, 3:1, 5:1, and 8:1 types of heart block (Figure 16.12).

Third-degree block Impulses fail to pass through the atrioventricular (AV) node and bundle of His, and the ventricle might start its own independent rhythm of beating, or you might see only the atria contracting.

After producing a complete (third-degree) heart block, click **Stop** and release the ligature by pulling on the release threads and observe whether a normal AV beat is restored.

Record your observations and answer the questions in the Laboratory Report.

8. All-or-None Law of the Heart

Increase the pressure on the AV ligature until the ventricle is completely quiet (no impulses are reaching the ventricle). Using electrodes, stimulate the ventricle to determine the threshold stimulus. Record the contraction height. Using a slow speed (50/sec), stimulate the ventricle with increasing voltages (single stimuli), recording the contraction height at each voltage.

Record your observations and answer the questions in the Laboratory Report. ◀

Copyright © 2015, 2011, 2008 Pearson Education, Inc.

Cardiac Function

Name _____

Date _____ Section _____

Score/Grade _____

Physiology of Amphibian or Reptilian Heart

1. Normal heartbeat
Place your record in the following space.

What is the heart rate at room temperature?_____ beats/min
Duration of systole = _____ sec   Duration of diastole = _____ sec
What causes the delay between the beat of the atria and the ventricle?

_____

_____

2. Refractory period of heart
Place recording of extra systole in the following space.

During which part of the ventricular cycle can extra systoles be obtained?

_____

_____

How does the duration of the ventricular refractory period compare with that for skeletal muscle? For a neuron?

_____

_____

Of what value is the length of the refractory period to the pumping action of the heart?

_____

_____

What causes the compensatory pause following an extra systole?

_____

_____

Copyright © 2015, 2011, 2008 Pearson Education, Inc.

3. Tetanization of heart

Place your record in the following space.

How do your results compare with tetanization of skeletal muscle?

_____

_____

How is this response related to the refractory period?

_____

_____

Explain the response obtained and its importance in the functioning of the heart as a pump.

_____

_____

4. Starling's law of the heart

Place your record in the following space.

Under what conditions would Starling's law be of importance in the intact animal?

_____

_____

A person suffering from heart failure often has an enlarged heart (hypertrophy). How is this hypertrophy related to Starling's law?

_____

_____

5. Temperature effects

Place your comparative records in the following space.

Cold Ringer's solution (HR = _____) Warm Ringer's solution (HR = _____)

How do you explain the changes seen when temperature is altered?

_____

_____

Copyright © 2015, 2011, 2008 Pearson Education, Inc.

6. Drug effects

Drug	Heart Rate	Contractility	Tone	Explanation
Acetylcholine				
Epinephrine				
Pilocarpine				
Nicotine				
Atropine				

Explain the mechanisms whereby acetylcholine and epinephrine alter the heart rate.

Acetylcholine:

_____

_____

Epinephrine:

_____

_____

Explain the response of the heart to acetylcholine following application of atropine.

_____

_____

7. Heart block

In the following space, place any interesting record that you obtained.

What types of heart block did you observe?

_____

_____

Exactly how does the ligature produce a heart block in this preparation?

_____

_____

How might a heart block be produced in pathological cases?

_____

_____

Explain how the excitation generated by the cardiac pacemaker spreads to the ventricular muscle fibers.

_____

_____

Copyright © 2015, 2011, 2008 Pearson Education, Inc.

8. All-or-none law of the heart

Place your record in the following space.

What is the all-or-none law of the heart?

_____

_____

How does the anatomical structure of the heart make this law possible?

_____

_____

How does your recording compare with that for skeletal muscle stimulated with increasing stimulus strengths? Why is there a difference in response?

_____

_____

APPLY WHAT YOU KNOW

1. Compare the response of skeletal muscle and cardiac muscle when both are infected with *Clostridium tetani*.

_____

_____

_____

_____

2. How would valvular stenosis (narrowing) of the atrioventricular valves affect ejection of blood from the heart?

_____

_____

_____

_____

Copyright © 2015, 2011, 2008 Pearson Education, Inc.

Human Cardiovascular Function

17

CHAPTER 17 INCLUDES:

 Vernier   2 Vernier® Activities

 PowerLab   2 PowerLab® Activities

BIOPAC Systems, Inc.   2 BIOPAC® Activities

PhysioEx 9.1

For more exercises on Human Cardiovascular Function, visit PhysioEx™ (www.physioex.com) and choose Exercise 5: Cardiovascular Dynamics.

OBJECTIVES

After completing this exercise, you should be able to

1. Explain the physiological basis of heart sounds.
2. Measure blood pressure and evaluate the effects of posture and sensory stimuli on blood pressure.
3. Record and explain the physiology of the peripheral pulse wave.
4. Record and explain the physiological basis of the electrocardiogram (ECG) and examine its use as a diagnostic tool.

Auscultation of Heart Sounds

Auscultation of the heart means listening to and studying the various sounds arising from the heart as it pumps blood. These sounds are the result of vibrations produced by the heart valves when they close. The heart sounds can be heard by placing the ear against the chest or by using a **stethoscope**. The vibrations producing the sounds can be visually displayed by using a heart sound microphone and physiological recorder to produce a **phonocardiogram**. Of the four major heart sounds, only the first two can be heard without using special amplification.

First heart sound   Produced at the beginning of systole when the atrioventricular (AV) valves close and the semilunar (SL) valves open. This sound has a low-pitched tone commonly termed the *lub* sound of the heartbeat.

Second heart sound   Occurs during the end of systole and is produced by the closure of the SL valves, the opening of the AV valves, and the resulting vibrations in the arteries and ventricles. Owing to the higher blood pressures in the arteries, the sound produced is higher pitched than the first heart sound. It is commonly referred to as the *dub* sound.

Third heart sound   Occurs during the rapid filling of the ventricles after the AV valves open and is probably produced by vibrations of the ventricular walls.

Fourth heart sound   Occurs at the time of atrial contraction and is probably due to the accelerated rush of blood into the ventricles.

► ACTIVITY 17.1
Auscultation of Heart Sounds
Materials

☐ 1 stethoscope

1. Using a stethoscope, listen to your partner's heart sounds, paying special attention to the four major **auscultatory areas** on the chest where the sounds from each valve can be heard most clearly (Figure 17.1).
2. If equipment is available, make a recording of your partner's phonocardiogram, using a heart sound microphone and recorder. Obtain readings of the heart sounds at each of the auscultatory areas and compare them for differences in vibratory patterns. ◄

Copyright © 2015, 2011, 2008 Pearson Education, Inc.

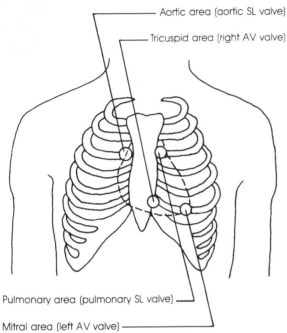

Aortic area (aortic SL valve)

Tricuspid area (right AV valve)

Pulmonary area (pulmonary SL valve)

Mitral area (left AV valve)

FIGURE 17.1 The four major auscultatory areas. SL, semilunar; AV, atrioventricular.

Blood Pressure

Blood pressure is one of the most useful clinical measurements that can be taken. Blood pressure refers to the pressure exerted by the blood against the vessel walls, the arterial blood pressure being the most useful, hence the most frequently measured, pressure. You should become familiar with the following pressures used in cardiovascular physiology.

Systolic blood pressure The highest pressure in the artery, produced in the heart's contraction (systolic) phase. The normal value for a 20-year-old man is 120 mm Hg.

Diastolic blood pressure The lowest pressure in the artery, produced in the heart's relaxation (diastolic) phase. The normal value for a 20-year-old man is 80 mm Hg.

Pulse pressure The difference between the systolic and diastolic pressures. The normal value is 40 mm Hg.

Mean blood pressure Diastolic pressure plus one-third of the pulse pressure. This is the average effective pressure forcing blood through the circulatory system. The normal value is 96–100 mm Hg.

The mean blood pressure is a function of two factors—cardiac output (CO) and total peripheral resistance (TPR). Peripheral resistance depends on the caliber (diameter) of the blood vessels and the viscosity of the blood.

Mean blood pressure = Cardiac output (ml/sec) × Total peripheral resistance (TPR units)

Cardiac output (ml/min) = Heart rate (beats/min) × Stroke volume (ml)

Thus, the measurement of blood pressure provides us with information on the heart's pumping efficiency and the condition of the systemic blood vessels. In general, we say that the systolic blood pressure indicates the force of contraction of the heart, whereas the diastolic blood pressure indicates the condition of the systemic blood vessels. (For instance, an increase in the systolic blood pressure and pulse pressure indicates a decrease in vessel elasticity.)

Blood pressure varies with a person's age, weight, and sex. Normal blood pressure is usually 90–119 mm Hg systolic and 60–79 mm Hg diastolic. Children tend to have lower blood pressure, and it usually increases as an individual ages. The increase in blood pressure with age is caused largely by the overall loss of vessel elasticity with age, part of which is due to the increased deposit of cholesterol and other lipids in the blood vessel walls. In addition to these factors, blood pressure values can change significantly over very short periods of time, so it is important to evaluate this parameter frequently to identify a possible pathology accurately.

STOP AND THINK

Although stressful events and the release of associated hormones such as epinephrine and cortisol can cause spikes in blood pressure, no connection has been established between chronic hypertension and stress. However, when our body is exposed to prolonged periods of stress, it is not surprising to find a significant effect on the cardiovascular system. Although stress itself is a risk factor for cardiovascular disease, stress can also exacerbate the effects of other preexisting risk factors such as high cholesterol and high blood pressure. Using your knowledge of the effects of epinephrine and cortisol, explain why reducing stress will help cardiovascular function.

Measurement of Blood Pressure

Blood pressure can be measured either directly or indirectly. In the **direct method**, a cannula is inserted into the artery and the direct head-on pressure of the blood is measured with a transducer or mercury manometer. In the **indirect method**, pressure is applied externally to the artery and the pressure is determined by listening to arterial sounds (using a stethoscope) below the point where the

Copyright © 2015, 2011, 2008 Pearson Education, Inc.

pressure is applied (Figure 17.2). This is called the **auscultatory method** because the detection of the sounds is termed auscultation. Pressure is applied to the artery, using an instrument called the **sphygmomanometer**. It consists of an inflatable rubber bag (cuff), a rubber bulb for introducing air into the cuff, and a mercury or aneroid manometer for measuring the pressure in the cuff. Human blood pressure is most commonly measured in the brachial artery of the upper arm. In addition to being a convenient place for taking measurements, it has the added advantage of being at approximately the same level as the heart, so pressures obtained closely approximate the pressure in the aorta leaving the heart. This allows us to correlate blood pressure with heart activity.

▶ ACTIVITY 17.2
Auscultatory Method Using a Stethoscope

Materials

☐ 1 stethoscope

☐ 1 sphygmomanometer

Have the subject sit with his or her arm resting on a table. Wrap the pressure cuff snugly around the bare upper arm, making certain that the inflatable

bag within the cuff is placed over the inside of the arm where it can exert pressure on the brachial artery. Wrap the end of the cuff around the arm and tuck it into the last turn or press the fasteners together to secure the cuff on the arm. Close the valve on the bulb by turning it clockwise.

Place the bell of the stethoscope below the cuff and over the brachial artery where it branches into the radial and ulnar arteries (Figure 17.2). Use your fingers, rather than your thumb, to hold the stethoscope over the artery; otherwise, you might be measuring your thumb arterial pressure rather than the subject's brachial artery pressure. With no air in the cuff, no sounds can be heard. Inflate the cuff so that the pressure is above diastolic (80–90 mm Hg), and you will be able to hear the spurting of blood through the partially occluded artery. Increase the cuff pressure to about 160 mm Hg; this pressure should be above normal systolic pressure so that the artery is completely collapsed and no sounds are heard.

Now, open the valve and gradually lower the pressure in the cuff. As the pressure decreases, you will be able to hear four phases of sound changes; these were first reported by Nikolai Korotkoff in 1905 and are called **Korotkoff sounds**.

Phase 1 Appearance of a fairly sharp thudding sound that increases in intensity during the next 10 mm Hg of drop in pressure. The pressure when the sound first appears is the **systolic pressure**.

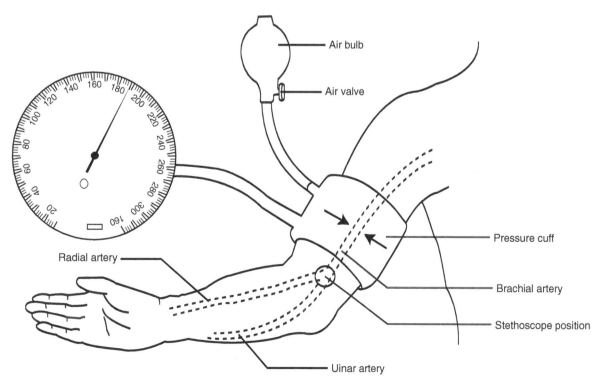

FIGURE 17.2 Apparatus for measuring blood pressure indirectly.

Copyright © 2015, 2011, 2008 Pearson Education, Inc.

Phase 2   The sounds become a softer murmur during the next 10–15 mm Hg of drop in pressure.

Phase 3   The sounds become louder again and have a sharper thudding quality during the next 10–15 mm Hg of drop in pressure.

Phase 4   The sounds suddenly become muffled and reduced in intensity. The pressure at this point is termed the **diastolic pressure**. This muffled sound continues for another drop in pressure of 5 mm Hg, after which all sound disappears. The point at which the sound ceases completely is called the **end diastolic pressure**. It is sometimes recorded along with the systolic and diastolic pressures in this manner: 120/80/75.

The auscultatory method has been found to be fairly close to the direct method in the pressures recorded: usually the systolic pressure is about 3–4 mm Hg lower than that obtained with the direct method.

Practice taking blood pressure on your partner until you become adept at detecting the systolic and diastolic sounds. You will find this can be quite difficult in some people, especially those whose arteries are located deep in the body tissues.

17.2 VERNIER® VERSION

Materials

☐ 1 Vernier® Lab Pro

☐ 1 blood pressure sensor

☐ 1 blood pressure cuff

Materials for this exercise will have been set up for you. If not, connect the blood pressure sensor to the Vernier® computer interface and the blood pressure cuff using the manufacturer's instructions. Open the "08 Blood Pressure Exercise" file from the Human Physiology with Vernier folder.

1. Have the subject sit with his or her arm resting on a table. Wrap the pressure cuff snugly around the bare upper arm, making certain that it is placed with the arrow labeled "Artery" positioned over the brachial artery.

2. Click **Collect** to begin data collection. Immediately begin to inflate the cuff until cuff pressure reaches at least 160 mm Hg. Stop pumping.

3. The software will calculate systolic, diastolic, and mean arterial pressure as the cuff automatically deflates. These values will be displayed on the computer screen. The program will stop calculating blood pressure when the readings stabilize. At this point, you can terminate data collection by clicking **Stop**. Release the pressure from the cuff but do not remove it.

4. Record the systolic, diastolic, and mean arterial pressure.

5. Repeat the recording, using the positions or stimuli described in Activities 17.3 and 17.4.

17.2 BIOPAC® VERSION

☐ 1 computer

☐ 1 BIOPAC® Student Lab software PC 3.7.0 or Mac 3.0.7 or greater

☐ 1 BIOPAC® data acquisition unit

☐ 1 blood pressure cuff (SS19L)

☐ 1 stethoscope (SS30L)

Materials for this experiment will have been set up for you.

Students should work in groups of three, with one student playing the role of the experimental subject, another the recorder, and another the director. It is important to allow only the recorder to view the computer screen while an experiment is being performed. The subject should not have had any history of a cardiovascular disorder. The role of the recorder is to start and stop the recording and to add markers to the recording. The director will call out the points of systolic and diastolic pressure. Use the director's name to identify the lesson file.

Start the BIOPAC® Student Lab Program and choose lesson L16-Bp-1, "Blood Pressure."

1. Plug the blood pressure cuff (SS19L) into channel 1 and the stethoscope (SS30L) into channel 3. In this experiment, you will not be using the electrode lead set (SS2L), although it should be plugged in to channel 4. Turn the data acquisition unit on.

2. Use alcohol swabs to clean the earpieces and the diaphragm of the stethoscope. Repeat the cleaning after use and prior to use with a new subject.

    Prior to calibrating the equipment, click **File**, choose **Preferences**, and select **Journal Preferences** in the pop-up window. Click **OK** and then choose **Show Minimal Journal Text** in the next window. Click **OK**.

3. Click **Calibrate** and follow the on-screen prompts. The blood pressure cuff is not on the subject while calibrating.

4. Director inflates the cuff to 100 mm Hg and informs the recorder.

5. Recorder clicks **OK**.

6. Director deflates cuff to 40 mm Hg and informs recorder.

7. Recorder clicks **OK** and tells the director to tap the stethoscope diaphragm twice. Calibration will stop in a few seconds.

Copyright © 2015, 2011, 2008 Pearson Education, Inc.

To obtain an optimal recording, the team should follow these guidelines:

a. The subject should not have consumed alcohol or caffeine or have smoked for an hour prior to the recording.
b. The arm being recorded should be in a completely relaxed position.
c. The equipment should be arranged to facilitate the comfort of the subject and allow ease of access to the cuff and the pressure dial.
d. The director should practice inflating the cuff and releasing the pressure so that the rate of decrease is constant at about 2–3 mm Hg per second.
e. The test must be conducted in a room quiet enough for sounds to be heard through the stethoscope.
f. Use the index and middle finger to feel the pulsations of the brachial artery on the inside of the elbow in the antecubital fossa; mark this area to facilitate the placement of the stethoscope diaphragm during the experiment.

Procedure

Have the subject sit with his or her arm resting on a table. Wrap the pressure cuff snugly around the bare upper arm (Figure 17.2), making certain that it is placed with the arrow labeled artery positioned over the brachial artery. The director will place the stethoscope earpieces in the subject's ear and control the release of pressure. He or she will also place the diaphragm on the location found earlier and listen for the Korotkoff sounds.

The recorder will control the computer and insert event markers by pressing **F9** (Windows) or **Esc** (Mac) when instructed by the director.

1. The recorder should click **Record** and follow the onscreen instructions.
2. The director should now inflate the cuff to about 160 mm Hg and inform the recorder, who should click **OK**. Once the recording starts, the recorder should insert an event marker.
3. The director should now release the pressure at a rate of 2–3 mm Hg/sec and call out when the first Korotkoff sound is heard.
4. The recorder should immediately insert the systolic event.
5. The director continues releasing pressure until the sounds disappear and immediately informs the recorder, who will insert the diastolic event marker. The recorder then clicks **Suspend** while the director deflates the cuff completely to avoid any discomfort to the subject.

The data should resemble that seen in Figure 17.3.

Click the first event marker and label it "Right/left Arm Sitting" in the marker region above the top data panel. Click the second marker and label this "Systole," and the third marker should be labeled "Diastole."

Repeat the recording by clicking **Resume**, using the positions or stimuli described in Activities 17.3 and 17.4. Label these recordings as you did above, using "Supine" and "Standing" for the first

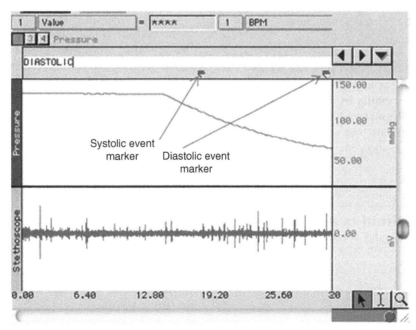

FIGURE 17.3 Blood pressure measurement using BIOPAC® software.

Copyright © 2015, 2011, 2008 Pearson Education, Inc.

marker in each recording. Click **Done** when all data from a single subject has been collected. After you do so, a prompt asks you to confirm that you have finished with all recordings. Click **Yes**, and the data will be written to file. If you need to repeat the procedure for another subject, make the appropriate choice in the next pop-up window. If not, you can proceed to data analysis.

Data Analysis
Select **Review Saved Data** from the **Lessons** menu. Select your data file. It will open and display your collected data. Three panels will be visible, labeled **Pressure, Stethoscope**, and **ECG**. We will be examining the data in the first two panels only. The measurement boxes are above the marker region in the data window. Each measurement has three sections: channel number, measurement type, and result.

Using the I-beam cursor, select the event marker labeled **Systole** and record the value in Channel 1. This is the systolic blood pressure. Do the same for the marker labeled **Diastole**. Using the familiar systolic pressure/diastolic pressure format (example 120/80), **enter the information in the table in your Laboratory Report**. Repeat this for the subsequent systolic and diastolic markers for each recording made. ◄

► ACTIVITY 17.3
Postural Effects on Blood Pressure

Materials

☐ 1 stethoscope
☐ 1 sphygmomanometer

Measure your partner's blood pressure while she is lying down (supine), sitting, and standing. **Record your results in the table in the Laboratory Report** and briefly explain the changes in pressure that accompany these changes in body position. ◄

► ACTIVITY 17.4
Sensory Stimulus and Blood Pressure: Cold Pressor Test

Materials

☐ 1 stethoscope
☐ 1 sphygmomanometer
☐ 1 small bucket of ice water

This test is used to demonstrate the effect of a sensory stimulus (cold) on blood pressure. A normal reflex response to a cold stimulus is an increase in blood pressure (both systolic and diastolic). In a normal individual, the systolic pressure will rise no more than 10 mm Hg, but in a hypertensive individual, the rise might be 30–40 mm Hg.

1. Have the subject sit down comfortably or lie supine.
2. Record the systolic and diastolic blood pressure every 5 min until a constant level is obtained.
3. Immerse the subject's free hand in ice water (approximately 5°C) to a depth well above the wrist.
4. After a lapse of 10–15 sec, obtain the blood pressure every 30 to 45 sec for up to 3 min. Allow at least 15 sec between readings.
5. How long did it take for the pressure to return to normal?

Does the blood pressure return to normal after immersion in ice water and, if so, how long does it take? Explain the physiological mechanisms operating in this experiment. ◄

Arterial Pulse Wave

The blood pressure within an artery varies during each cardiac cycle. The highest pressure (systolic) occurs when the ventricle contracts to force blood into the artery; the lowest pressure (diastolic) occurs when the heart is in its relaxation phase and no blood is flowing through the semilunar valves. The difference between the systolic and diastolic pressures is called the **pulse pressure**. A recording of these pressure changes in an artery during one cycle of the heart is called an arterial **pulse wave**. A normal pulse wave over the aorta is shown in Figure 17.4. The **dicrotic notch** results when the aortic semilunar valves close, causing the blood in the aorta to rebound against the arterial walls to produce a slight elevation in pressure.

The magnitude and contour of the arterial pulse wave are directly related to the stroke volume and inversely related to the compliance (elasticity) of the arterial vessels. As the vessels lose their compliance (as with age or in arteriosclerosis), the stroke volume increases, and the height of the pulse wave increases (pulse pressure increases). Thus, an examination of the pulse wave can give

Copyright © 2015, 2011, 2008 Pearson Education, Inc.

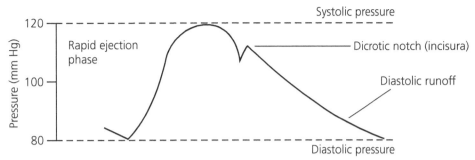

FIGURE 17.4   Typical arterial pulse wave.

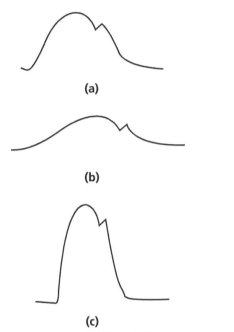

FIGURE 17.5   Pulse waves. (a) Normal. (b) In aortic stenosis. (c) In arteriosclerosis.

TABLE 17.1   Change in Velocity of Pulse Wave with Age

Age (Yr)	Pulse Wave Velocity (m/sec)
5	5.2
20	6.2
40	7.2
80	8.3

▶ ACTIVITY 17.5

Recording the Peripheral Pulse Using the Physiograph

Materials

☐ 1 Physiograph™ with recording paper

☐ 1 photoelectric pulse transducer (plethysmograph)

In this activity, you will not record the pulse wave over an artery but from the tip of the finger; this is called a peripheral pulse. It is recorded using a photoelectric pulse transducer, which measures changes in blood volume (plethysmography). A light source in the transducer transilluminates the fingertip, and a photoconductor detects changes in light intensity within the finger caused by pulsatile variations in blood volume.

1. With the subject seated, attach the transducer snugly to the palmar surface of the middle finger. Record the pulse for 10 sec with the subject's arm resting on the lab table.

    Now have the subject raise the transducer above the head (arm extended) for 30 sec and record the pulse during the last 10 sec.

    Then have the subject lower the transducer (arm hanging at the side) for 30 sec and record during the last 10 sec.

valuable clues to the functioning of the arteries and heart, as is seen in the abnormal waves pictured in Figure 17.5.

The velocity of the pulse wave as it travels down the artery is also an important clinical measurement. The arterial pulse wave moves over the large arteries at a rate of 3–5 m/sec and over the small arteries at 14–15 m/sec. The difference in velocity is related to the compliance of the vessels—the less compliance a vessel has, the faster the pulse wave will move over it (as in the small arteries). Thus, a measurement of the velocity of the pulse wave can also provide useful information about changes in the vessel's elasticity (compliance). The change in vessel elasticity with age is apparent if you examine the velocity of the pulse wave over the aorta at various stages (Table 17.1).

Copyright © 2015, 2011, 2008 Pearson Education, Inc.

2. Valsalva maneuver

    Note: A person with a history of cardio-vascular disease should not be used as a subject for this experiment.

    With the subject's arm resting on the table, record the pulse for 10 sec. Then have the subject inhale as deeply as possible, hold his or her breath, and exert as much internal abdominal pressure as possible, while you record the pulse. Continue recording for about 15 sec while performing the maneuver, exhale, and then record for 15 more seconds. Compare the heart rate and the height of each wave just prior to inhalation with the rate just prior to exhalation and a few seconds after exhalation. **Record these values under Pulse Valsalva 1, 2, and 3, respectively, in the table in the Laboratory Report.**

    Attempting to exhale forcefully against a closed glottis is called the Valsalva maneuver. It is commonly performed during forceful defecation or when lifting heavy weights. The contraction of the internal intercostal and abdominal muscles during this maneuver greatly increases the intrathoracic and intra-abdominal pressures, which impedes the venous return of blood to the heart.

3. After recovery from the Valsalva maneuver, wrap a sphygmomanometer cuff around the upper arm and record the peripheral pulse while the cuff is inflated to occlude the brachial artery. Continue recording the pulse as the cuff is slowly deflated at a rate of 5 mm Hg/sec. Record the height of the pulse and the rate for when the cuff is inflated, while reducing pressure, and when all pressure is released.

4. **Enter these values in the table in your Laboratory Report.**

17.5 POWERLAB® VERSION

Materials (per group)

☐ 1 computer

☐ 1 PowerLab® data acquisition unit

☐ 1 pulse transducer MLT1010/D

Materials and equipment for this activity will have been set up for you. Make sure that the finger pulse transducer is connected to the Channel 2 input on your PowerLab® system.

1. For this investigation, launch **Chart** by double-clicking the **Volume pulse settings** file usually located in the EKG And Peripheral Circ/Settings Files folder or on the desktop of your computer. A two-channel Chart window will appear. Channel 1, labeled volume pulse, is a computation (the time integral) based on data in Channel 2. Raw data will be displayed in Channel 2. The Channel 2 waveforms should resemble the image in Figure 17.4 and the lower panel in Figure 17.9.

2. With the subject seated, attach the finger pulse transducer snugly to the palmar surface of the middle finger. Click **Start** and record the pulse for 10 sec with the subject's arm resting on the lab table. Click **Stop**.

3. Click **Commands** and choose **Add Comment**. Type in the arm position or other suitable information in the **Add Comment** window to label your record.

4. Now have the subject raise the transducer above the head (arm extended) for 30 sec and then click **Start**. Record the pulse for 10 sec, as before, and click **Stop**. Have the subject lower the transducer (arm hanging at the side) for 30 sec and, once again, record the last 10 sec. Add a comment to your records as described before.

5. Valsalva maneuver

    Note: A person with a history of cardio-vascular disease should not be used as a subject for this experiment.

    Click **Start**. With the subject's arm resting on the table, record the pulse for 10 sec and then have the subject inhale as deeply as possible, hold his or her breath, and exert as much internal abdominal pressure as possible while you continue to record the pulse for at least 15 sec. Have the subject exhale slowly, and continue recording for 15 sec after exhaling. Click **Stop**. Compare the heart rate, and the height of each wave just prior to inhalation, with the rate just prior to exhalation and a few seconds after exhalation. **Record these values under Pulse Valsalva 1, 2, and 3, respectively, in the table in the Laboratory Report.**

    Attempting to exhale forcefully against a closed glottis is called the Valsalva maneuver. It is commonly performed during forceful defecation or when lifting heavy weights. The contraction of the internal intercostal and abdominal muscles during this maneuver greatly increases the intrathoracic and intra-abdominal pressures, which impedes the venous return of blood to the heart.

Copyright © 2015, 2011, 2008 Pearson Education, Inc.

EXPLAIN THIS!

Why does the Valsalva maneuver impede venous return of blood to the heart? [Your explanation must go beyond stating that an increase in intrathoracic and intra-abdominal pressures reduces blood flow back to the heart.]

---

6. After recovery from the Valsalva maneuver, wrap a sphygmomanometer cuff around the upper arm and record the peripheral pulse while the cuff is inflated to occlude the brachial artery. Continue recording the pulse as the cuff is slowly deflated at a rate of 5 mm Hg/sec. Record the height of the pulse and the rate for when the cuff is inflated, while reducing pressure, and when all pressure is released. **Record these values in the table in your Laboratory Report.** ◀

CLINICAL APPLICATION

Cardiac Arrest

We have all seen or heard about the suddenness and significant danger associated with cardiac arrest, where the heart ceases to pump blood effectively to our tissues. Cardiac arrest can be caused by the occlusion of a coronary artery where blood supply to the heart is reduced or blocked, causing cell damage or death (infarction) in a matter of minutes. Other causes can include toxins, hypovolemia (a drop in blood volume), the accumulation of fluid within the pericardial sac (cardiac tamponade), and hypokalemia or hyperkalemia (lower- or higher-than-normal levels of potassium). Intervention in the form of cardiopulmonary resuscitation (CPR) and defibrillation enhance the chance of survival. CPR is a combination of chest compressions and rescue breathing techniques meant to maintain the supply of oxygen to the tissues. It allows a person to buy time. If the ventricles are defibrillating or are in a state of tachycardia, defibrillation of the heart musculature can help restore heart function. A defibrillator will deliver a therapeutic dose of electricity to the heart, depolarizing a critical mass of cardiac muscle, simultaneously terminating the arrhythmia and restoring a sinus pattern of contraction. In the absence of any heart function, pharmacological intervention is usually used to attempt to restore heart contractions.

Electrocardiogram

Every living cardiac cell undergoes a regular sequence of electrical changes that initiate the contractile activity (systole) and the relaxation (diastole) of the cell. Thus, the contraction of the heart is associated with a compound action potential that is initiated at the sinus node and sweeps over the conduction path of the heart, preceding the mechanical contraction of the cardiac fibers. During this depolarization and repolarization of the myocardium, a potential difference is created between different regions on the surface of the heart. A separation of charge or potential difference is called a **dipole**. The electrical potential of the dipole is conducted through an electrolyte solution, such as the interstitial fluid and blood plasma, and eventually reaches the surface of the skin. By placing electrodes on the skin surface, we are able to detect and record the electrical activity over the heart surface prior to its contraction. By measuring the potential changes in various directions across the heart, it is possible to detect abnormalities.

The **electrocardiogram (EKG or ECG)** is a graphic record of the action potentials of the heart. It is recorded with an **electrocardiograph**, and the study of this cardiac electrical activity is called **electrocardiography**.

Electrocardiograph

The instrument that amplifies and records the heart's action potentials is actually a galvanometer, a device used by electricians to measure the passing of an electric current. Several types have been employed: the early string galvanometer used by Willem Einthoven, the electronic polygraph recorder, the cathode ray oscilloscope, and radiotelemetry recorders such as those used in space physiology research. The PowerLab® and Vernier® data acquisition systems collect and display the EKG on the computer screen.

Electrodes

The limb electrodes generally used are slightly concave metal plates designed to fit snugly over the wrists and ankles. The chest electrode is a flattened disk. Because the skin has a high resistance, an electrolyte jelly (NaCl or KCl) in an abrasive base is first rubbed on the skin to remove the oil and dead cells and to form a conducting surface between the skin and the metal electrode, thus improving conduction of the impulse.

Newer self-adhesive tab electrodes are available from a number of manufacturers. The PowerLab®, Vernier®, and BIOPAC® systems use snap connectors or alligator clips for the EKG leads.

Standard Limb Leads

Various types of electrode positions can be used to record an EKG. The particular arrangement of the two recording electrodes is called a **lead**. The

Copyright © 2015, 2011, 2008 Pearson Education, Inc.

relative position of the two electrodes influences the direction and amplitude of the potentials recorded. To standardize the procedure so that results from different laboratories can be compared and evaluated, cardiologists have agreed on certain conventional requirements for the recording of the EKG.

The most common recording procedure is to use the standard limb leads. Electrodes are placed on the left arm, right arm, left leg, and right leg. The electrode on the right leg is a ground connection that prevents unwanted external potential fields from distorting the record. The other three electrodes are used in pairs to detect the cardiac potential. The electrocardiograph is calibrated so that 1 cm of vertical deflection represents 1 mV of potential difference, and the standard paper speed used is 25 mm/sec.

Einthoven's Triangle and Law

Willem Einthoven, the father of electrocardiography, originated many of the conventions used in the recording of electrocardiograms. He visualized the three standard limb leads enclosing the heart in a triangle, often referred to as Einthoven's triangle (Figure 17.6). Einthoven also found a relationship between the amplitude of the QRS complexes in each lead, such that lead I + lead III = lead II (Einthoven's law).

Lead I: Right arm to left arm   The right arm is connected to the negative terminal of the electrocardiograph and the left arm to the positive terminal. When the right arm is negative to the left arm, the record shows an upward deflection. Thus, lead I measures the potential difference between the electrodes on the left and right arms, across the base of the heart.

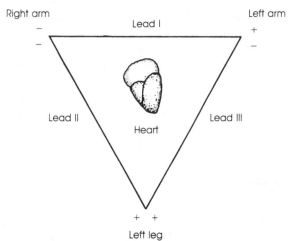

FIGURE 17.6   Einthoven's triangle.

Lead II: Right arm to left leg   The right arm is connected to the negative terminal and the left leg to the positive terminal. Thus, lead II measures the potential difference between the left leg and the right arm, along the long axis of the heart from base to apex.

Lead III: Left arm to left leg   The left arm is connected to the negative terminal and the left leg to the positive terminal. This combination allows lead III to measure the potential difference between the left leg and the left arm, along the left side of the heart.

The sinoatrial (SA) node initiates the cardiac impulse (epicardium in this area becomes negative first), and this wave of negativity sweeps over the heart. Because the SA node is nearer to the right arm, this area becomes negative while the left arm and left leg are still positive, and the deflection of the record is upward in those leads (I and II). The left arm is closer to the SA node, so in lead III the first deflection is also upward as the left arm becomes negative in reference to the left leg.

Components of a Normal EKG Complex

A normal EKG for a single cardiac cycle is shown in Figure 17.7.

> **P Wave**   Represents the spread of electrical activity (wave of negativity) over the atria after the initial depolarization of the SA node.
>
> **QRS Complex**   Represents the spread of the negativity wave (depolarization) through the ventricular musculature (Figure 17.7). A small amount of atrial repolarization also occurs at the same time.
>
> **PR Interval**   Time from the beginning of the P wave to the beginning of the QRS complex; interval between activation of the SA node and the beginning of ventricular depolarization. Any abnormal lengthening of this interval suggests some interference with conduction of the impulse through the atria, atrioventricular (AV) node, bundle of His, and Purkinje fibers.
>
> **T Wave**   Represents the repolarization of the ventricular musculature. It is of longer duration and lower amplitude than the depolarization wave (QRS complex), which indicates that the ventricular repolarization process is less synchronized and slower than the depolarization process.
>
> **QT Interval**   Represents the time from the beginning of the QRS complex to the end of the T wave, that is, from the beginning of ventricular depolarization to the end of ventricular repolarization. The QT interval varies with the heart rate, becoming shorter as the heart rate increases.

Copyright © 2015, 2011, 2008 Pearson Education, Inc.

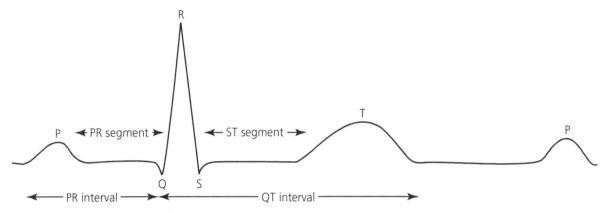

FIGURE 17.7   Normal electrocardiogram.

PR Segment   From the end of the P wave to the beginning of the QRS complex. During this time, the impulse is traveling through the AV node, AV bundle, and Purkinje fibers. These structures are within the heart myocardium; therefore, during this time, there is no change in the negativity of the surface of the heart, and we say that the record is isoelectric. (No change in potential is occurring.)

ST Segment   From the end of the S wave to the beginning of the T wave. During this time, the heart is completely depolarized, and therefore the record is again isoelectric. The position and shape of the ST segment are important in diagnosis.

Normal values for the duration and, in some cases, the voltage of the different phases of the EKG complex are shown in Table 17.2.

▶ ACTIVITY 17.6
EKG Using Physiograph™

Materials

☐ 1 Physiograph™

☐ 1 EKG electrode lead set

☐ 1 plethysmograph

1. Check the electrocardiograph's calibration so that 1 cm = 1 mV of potential difference on the vertical axis. Run a length of record at the standard speed of 25 mm/sec, using the settings for lead I in Table 17.3. Check the record for artifacts. Repeat, using the settings for leads II and III.
2. Check the amplitudes of the QRS complexes in leads I, II, and III. Do they obey Einthoven's law?

TABLE 17.2   Normal Values for Duration (and Voltage) of Different Phases of EKG Complex

Phase of Complex	Duration (sec) (Voltage [mv])
P wave	0.1 (0.2)
QRS complex (lead II)	0.08–0.12 (1)
T wave	0.16–0.27 (0.2–0.3)
PR interval	0.13–0.16
QT interval	0.3–0.34
PR segment	0.03–0.06
ST segment	0.08

3. Examine several inches of the record. Does the cycle length ever vary (arrhythmia)? Is there a change in cycle length (heart rate) with inspiration or expiration? Are any of the waves abnormal?
4. Make routine measurements of the following:

   PR interval
   P wave amplitude and duration
   QRS interval
   QT interval
   T wave amplitude and duration
   PR segment
   ST segment

TABLE 17.3   Bipolar Limb Leads for the EKG

	Left Wrist	Right Wrist	Left Ankle	Right Ankle
Lead I	Positive	Negative		Ground
Lead II		Negative	Positive	Ground
Lead III	Negative		Positive	Ground

Copyright © 2015, 2011, 2008 Pearson Education, Inc.

A PR interval greater than 0.2 sec is abnormal and indicates first-degree heart block. In second-degree heart block, there are P waves that are not followed by QRS waves; this might occur regularly or irregularly. Third-degree heart block is a complete AV dissociation in which P waves occur quite regularly but are not related to R waves.

The normal duration of the QRS complex is 0.08–0.12 sec. A duration of more than 0.12 sec indicates bundle branch block or that the beat has arisen in one of the ventricles—a so-called ventricular beat or extra systole.

Variations in the T wave are quite numerous and require an expert cardiologist for proper diagnosis. Inversion of the T wave is not abnormal, especially in lead III. Notching or splintering of the R wave (QRS) can be due to sudden changes in the electrical axis of the heart. Elevation of the ST segment by more than 2 mm is associated with acute injury or anoxia.

5. Have the subject exercise by doing several deep knee bends and record the EKG following the exercise. Are there any alterations in the patterns of the various waves or variations in the heart rate?

6. If the equipment is available, obtain a simultaneous recording of the EKG, phonocardiogram (heart sounds), and arterial pulse wave (using the finger or the radial artery). Compare the record obtained with that shown in Figure 17.8, a record of these three parameters during the cardiac cycle. The pressure pulse will be displaced to the right of the ventricular pressure curve shown because you will be measuring the pulse farther from the heart. Can you explain the relationships between the various events in the cardiac cycle?

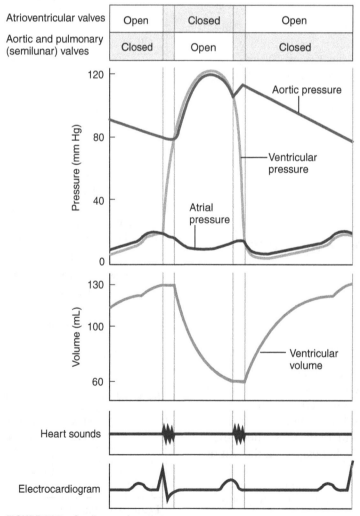

FIGURE 17.8 Cardiac cycle events.

Copyright © 2015, 2011, 2008 Pearson Education, Inc.

17.6 VERNIER® VERSION

Materials

☐ 1 computer

☐ 1 Vernier® Lab Pro

☐ 1 EKG sensor

☐ Disposable EKG tab electrodes (3 per student tested)

☐ Printer

The EKG sensor used for this exercise should be connected to the Vernier® computer interface. Open the Analyzing Heart EKG file from the Human Physiology with Vernier folder. Make sure that the tab electrodes are pressed firmly into place on the left and right wrists and the left ankle. Have the subject sit still, avoiding unnecessary movements, because they can interfere with data.

1. Using the settings for lead I in Table 17.3, attach the alligator clips to the electrode tabs. Click **Collect** to record the EKG.
2. Click **File** and select **Print** to print out a record of your recording.
3. Repeat the procedure, using the settings for leads II and III in Table 17.3, making sure that each is labeled appropriately.
4. Check the amplitudes of the QRS complexes in leads I, II, and III. Do they obey Einthoven's law?
5. Examine several seconds of each record. Does the cycle length ever vary (arrhythmia)? Is there a change in cycle length (heart rate) with inspiration or expiration? Are any of the waves abnormal?
6. Make routine measurements of the following:

   PR interval

   P wave amplitude and duration

   QRS interval

   QT interval

   T wave amplitude and duration

   PR segment

   ST segment

   A PR interval greater than 0.2 sec is abnormal and indicates first-degree heart block. In second-degree heart block, there are P waves that are not followed by QRS waves; this can occur regularly or irregularly. Third-degree heart block is a complete AV dissociation in which P waves occur quite regularly but are not related to R waves.

   The normal duration of the QRS complex is 0.08–0.12 sec. A duration of more than 0.12 sec indicates bundle branch block or that the beat has arisen in one of the ventricles—a so-called ventricular beat or extra systole. Variations in the T wave are quite numerous and require an expert cardiologist for proper diagnosis. Inversion of the T wave is not abnormal, especially in lead III. Notching or splintering of the R wave (QRS) can be due to sudden changes in the electrical axis of the heart. Elevation of the ST segment by more than 2 mm is associated with acute injury or anoxia.

7. Have the subject exercise by doing several deep knee bends and record the EKG following the exercise. Are there any alterations in the patterns of the various waves or variations in the heart rate?
8. If the equipment is available, obtain a simultaneous recording of the EKG, phonocardiogram, (heart sounds), and arterial pulse wave (using the finger or the radial artery). Compare the record obtained with that shown in Figure 17.8, a record of these three parameters during the cardiac cycle. The pressure pulse will be displaced to the right of the ventricular pressure curve shown because you will be measuring the pulse farther from the heart. Can you explain the relationships between the various events in the cardiac cycle?

17.6 POWERLAB® VERSION

Materials

☐ 1 computer

☐ 1 PowerLab® data acquisition unit

☐ Disposable EKG electrodes (3 per student tested)

☐ 3 lead shielded bio amp cable

☐ Snap-on shielded lead wires

☐ Printer

Materials and equipment for this experiment will have been set up for you. Launch **Chart** by double-clicking the **EKG Settings** file usually located in the EKG and Heart Sounds/Settings Files folder or on the desktop of your computer. This file will load the settings for the five-lead shielded bio amp cable used to record the EKG. Make sure that snap-tab electrodes are pressed firmly into place on the left and right wrists and the left ankle. Have the subject sit still, avoiding unnecessary movements, because they can interfere with data collection.

1. Using the settings for lead I in Table 17.3, click **Start** to record the EKG. Click **Stop** after you have obtained a 20-sec record. Use the comments window to label your recording of lead I.

Copyright © 2015, 2011, 2008 Pearson Education, Inc.

2. Select a portion of the record and print it out by pressing **File** and selecting **Print Selection**.
3. Repeat the procedure, using the settings for leads II and III in Table 17.3, making sure that each is labeled appropriately.
4. Check the amplitudes of the QRS complexes in leads I, II, and III. Do they obey Einthoven's law?
5. Examine several seconds of each record. Does the cycle length ever vary (arrhythmia)? Is there a change in cycle length (heart rate) with inspiration or expiration? Are any of the waves abnormal?
6. Make routine measurements of the following:

   PR interval

   P wave amplitude and duration

   QRS interval

   QT interval

   T wave amplitude and duration

   PR segment

   ST segment

A PR interval greater than 0.2 sec is abnormal and indicates first-degree heart block. In second-degree heart block, there are P waves that are not followed by QRS waves; this can occur regularly or irregularly. Third-degree heart block is a complete AV dissociation in which P waves occur quite regularly but are not related to R waves.

The normal duration of the QRS complex is 0.08–0.12 sec. A duration of more than 0.12 sec indicates bundle branch block or that the beat has arisen in one of the ventricles—a so-called ventricular beat or extra systole.

Variations in the T wave are quite numerous and require an expert cardiologist for proper diagnosis. Inversion of the T wave is not abnormal, especially in lead III. Notching or splintering of the R wave (QRS) might be due to sudden changes in the electrical axis of the heart. Elevation of the ST segment by more than 2 mm is associated with acute injury or anoxia.

7. Have the subject exercise by doing several deep knee bends and record the EKG following the exercise. Are there any alterations in the patterns of the various waves or variations in the heart rate?
8. Use the finger pulse transducer to obtain a simultaneous recording of the EKG and the arterial pulse wave. Compare the record obtained with that shown in Figure 17.8, a record of these two parameters (plus the phonocardiogram) during the cardiac cycle. The pressure pulse will be displaced to the right of the ventricular pressure curve shown because you will be measuring the pulse farther from the heart. Can you explain the relationships between the various events in the cardiac cycle?

17.6 BIOPAC® VERSION

Materials (per group)

☐ 1 computer

☐ 1 BIOPAC® student lab software PC 3.7.0 or Mac 3.0.7 or greater

☐ 1 BIOPAC® data acquisition unit

☐ 1 BIOPAC® electrode lead set (SS2L) connected to channel 1

☐ 1 BIOPAC® pulse plethysmograph (SS4LA or SS4L) connected to channel 2

☐ 4 BIOPAC® disposable vinyl electrodes (EL503) per subject

☐ 1 BIOPAC® stethoscope (SS30L)

Materials needed for this experiment will have been sent up for you. It is recommended that you work in pairs, taking turns being the subject and the recorder. In preparation for the experiment, the subject should remove all metal objects from his or her arms and legs.

Put a vinyl electrode slightly above the right wrist, above the left wrist, and on the medial surfaces of the right and left legs, just above the ankle. Make sure that these are pressed on firmly. Wait a few minutes for optimum electrode adhesion. Attach the leads, using the guidelines for lead I in Table 17.3.

Wrap the pulse transducer with the window facing the fleshy portion of the right index finger. Use the strap to create a snug fit.

Start the BIOPAC® student lab program and open lesson L07-ECG&P-1 ECG and Pulse. Type in a unique file name to store the data collected when instructed.

Prior to calibrating the equipment, click **File**, choose **Preferences**, and select **Journal Preferences** in the pop-up window. Click **OK** and then choose **Show Minimal Journal Text** in the next window. Click **OK.**

When the subject is seated, relaxed, and keeping still, click **Calibrate** and wait for the procedure to stop.

1. Click **Record** to start collecting data. The recorder should immediately insert an event marker by pressing **F9** (Windows) or **Esc** (Mac).
2. Collect data for 15 sec and then click **Suspend**. Your screen should resemble Figure 17.9.

Copyright © 2015, 2011, 2008 Pearson Education, Inc.

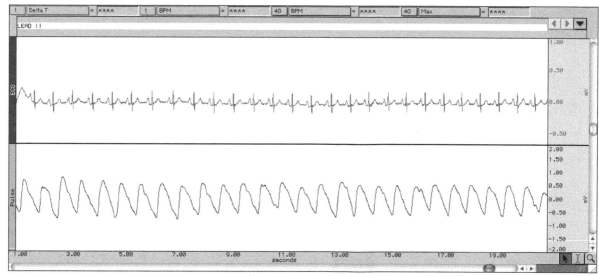

FIGURE 17.9   ECG and Pulse waves, using BIOPAC® software.

3. Repeat steps 1 and 2 using the electrode settings for leads II and III in Table 17.3. Click **Resume/Continue** to continue recording for each lead. Remember to insert an event marker prior to recording each lead.

4. Click the first event marker and label it "EKG Lead I" in the marker region above the top data panel. Click the second marker and label this "EKG Lead II." The third marker should be labeled "EKG Lead III."

Reconnect the electrodes using lead II. Have the subject exercise by doing several deep knee bends and record the EKG following the exercise. Label this recording "EKG Exercise Lead II." Are there any alterations in the patterns of the various waves or variations in the heart rate? **Enter your observations and answer the questions in the Laboratory Report.**

For the next set of tests, you will be using the data collected and displayed primarily in the lower panel.

1. Click **Continue/Resume** to start collecting data. The recorder should immediately insert an event marker.

2. Collect data for 15 sec and then click **Suspend**. Your screen should resemble Figure 17.3 (page 189). Click the event marker and label it "Pulse Seated."

3. Now have the subject raise the transducer above the head (arm extended) for 30 sec and then click **Continue/Resume**. Insert an event marker and record the pulse for 10 sec as before and then click **Suspend**. Insert another marker, have the subject lower the transducer (arm hanging at the side) for 30 sec, and,

once again, record the last 10 sec. Label these markers "Pulse Arm Raised" and "Pulse Arm to the Side," respectively.

4. Valsalva maneuver
Note: A person with a history of cardiovascular disease should not be used as a subject for this experiment. Click **Continue/Resume**. Insert a marker. With the subject's arm resting on the table, record the pulse for 10 sec. Then have the subject inhale as deeply as possible, hold his breath, and exert as much internal abdominal pressure as possible while you continue to record the pulse for at least 15 sec. Have the subject exhale slowly, and continue recording for 15 sec after exhaling. Click **Suspend**. Label the marker "Pulse Valsalva."

Compare the heart rate just prior to inhalation with the rate just prior to exhalation by counting the number of pulse waves in a set amount of time. **Record these values and enter the information in the Laboratory Report.**

Attempting to exhale forcefully against a closed glottis is called the Valsalva maneuver. It is commonly performed during forceful defecation or when lifting heavy weights. The contraction of the internal intercostal and abdominal muscles during this maneuver greatly increases the intrathoracic and intra-abdominal pressures, which impedes the venous return of blood to the heart.

5. After recovery from the Valsalva maneuver, wrap a blood pressure cuff around the upper arm and inflate to a pressure of 160 mm Hg.

Copyright © 2015, 2011, 2008 Pearson Education, Inc.

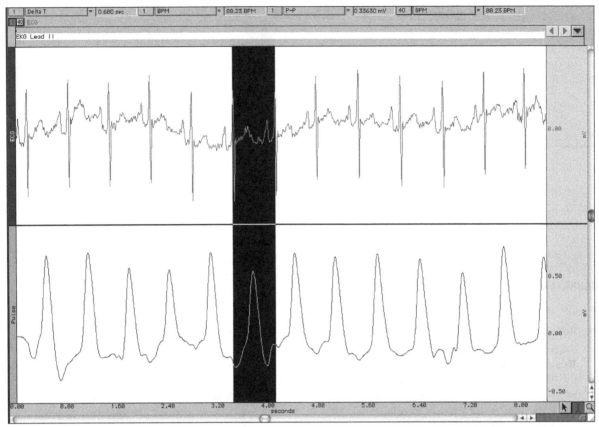

FIGURE 17.10 Measuring distance between peaks, using the I-beam tool in BIOPAC® software.

Click **Resume** to record the peripheral pulse. Insert a marker and continue recording the pulse as the cuff is slowly deflated at a rate of 5 mm Hg/sec. Continue recording until the pressure drops below 60 mm Hg and then click **Suspend**. Label the marker "Pulse with Cuff."

6. Click **Done/Yes/Analyze Current data file** and then **Yes** to save your data and open the data analysis feature.

Data Analysis

You should have nine event markers associated with this recording. You will be analyzing each of the events separately. Use the **zoom tool** to display approximately four cardiac cycles only to help you make your measurements. There are four measurement boxes displayed above the data panels. The measurement boxes will be labeled:

Channel	Measurement
CH 1	ΔT (time interval)
CH 1	BPM (rate)
CH 1	p-p (max–min) in selected area of channel 1
CH 40	p-p (max-min) in selected area of channel 40

1. Use the I-beam tool as shown in Figure 17.10 to select the areas described in the text under "Components of a Normal EKG Complex" (page 194) and in Figure 17.7 to identify and measure the following segments in the upper panel labeled "ECG":

   PR interval

   P wave amplitude and duration

   QRS interval

   QT interval

   T wave amplitude and duration

   PR segment

   ST segment

The values will be displayed alongside the CH 1 ΔT measurement boxes. **Enter these values in the table in the Laboratory Report. Repeat this step for recordings associated with event markers 1–4.**

2. Click the first CH 1 box and use the drop-down menu to choose **CH 40**. Choose **Delta T** in the associated drop-down measurement box.

3. Click the second CH 1 box, associated with BPM. Use the drop-down menu and choose

Copyright © 2015, 2011, 2008 Pearson Education, Inc.

CH **40**. Choose **BPM** in the associated drop-down measurement box.

4. Click the third CH 1 box, associated with (p-p). Use the drop-down menu to choose CH **40**. Choose **Max** as the associated drop-down measurement box.

5. Using the I-beam tool (Figure 17.10), select the area between two adjacent peaks in the lower panel labeled "Pulse." Try to go from peak to peak as precisely as possible. **Record the values for Delta T, BPM, and Max and enter these in your Laboratory Report.** ◀

INQUIRY-BASED ACTIVITY

Appendix C describes the format of a typical Laboratory Report. It is mandatory that you read the Appendix before you start planning your experiment.

Use any of the activities described previously in this chapter to design your own experiment. The independent variable you choose should be easily manipulated and appropriate for the activity concerned. Please also make sure that, if using a unique measure of the dependent variables, you establish the reliability of the measuring techniques. It is suggested that you use the same measuring techniques described in the activities.

Complete this checklist to make sure that you have covered all bases before you start your experiment:

What is the question you seek to answer?

Frame it in the form of a hypothesis.

What is the independent variable or treatment?

How will the independent variable or treatment be measured?

What is the control treatment?

How will you replicate your experiment?

How will you ensure that your subjects are similar enough to not introduce some other independent variability? Are there any standardized variables?

What is/are the dependent variable(s)?

How will the dependent variable(s) be measured?

What are your predictions?

Construct a table to record your observations easily.

How will you present the data collected graphically?

How will you analyze the data collected?

How will you know if the differences between the treated and the untreated samples are statistically significant?

Your laboratory instructor will describe how your Laboratory Report will be written.

Electrical Axis of the Heart

Using Einthoven's triangle and law, you can calculate the overall direction and magnitude of the electrical impulses conducted over the heart; this is called the **electrical axis** of the heart. The electrical axis is a valuable tool for clinical diagnosis because it indicates the approximate position the heart occupies in the thoracic cavity and possible hypertrophy of the heart chambers.

▶ ACTIVITY 17.7
Electrical Axis of the Heart

Use the printouts of the recordings obtained for each of the three EKG leads to complete this exercise. You can also use the measurement tools and the saved data from the BIOPAC®, Vernier®, or PowerLab® EKG activities to make these measurements.

1. Measure the height of the QRS deflections in each of the standard limb leads (I, II, and III). This height is usually measured in millimeters and converted to millivolts or microvolts. The height of the downward deflection is subtracted from the upward deflection height. Figure 17.11 shows examples.

2. The resultant deflections are plotted on the Einthoven triangle as vectors (Figure 17.12), appropriate units being used to represent the millimeter or millivolt height of each deflection. Each vector is drawn along the side of the triangle corresponding to the lead it represents. The vector is drawn from the center of each side and toward the (+) pole or electrode. If Einthoven's law is valid, perpendicular lines dropped from the point of each vector should meet a single point inside the

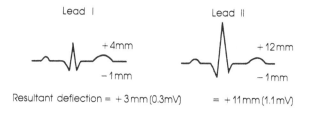

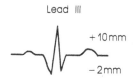

FIGURE 17.11 QRS deflections.

Copyright © 2015, 2011, 2008 Pearson Education, Inc.

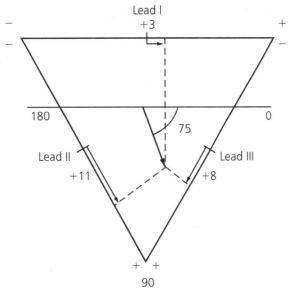

FIGURE 17.12   Electrical axis of the heart.

triangle. A line drawn from this intersection to the center of the triangle represents the electrical axis of the heart. The position of this axis is measured in degrees from the horizontal line drawn through the center of the triangle (75° in Figure 17.12). The length of the axis vector is a measure of the overall electrical potential of the heart in the axis direction.

Normally, the heart has an electrical axis of 59°. If the axis is less than 0°, it is termed a **left axis deviation**. If it is greater than 90°, it is termed a **right axis deviation**. In some cases, the heart itself can be pushed out of its normal position, producing a left or right shift in the axis deviation (as with pregnancy, excessive weight, or abdominal tumors). A left axis deviation can also be caused by hypertrophy of the left ventricle, as with systemic hypertension. A right axis deviation can be produced by right ventricle hypertrophy, as with pulmonary hypertension.

3. Calculate the electrical axis of the subject's heart and compare it with the normal expected deviation. ◄

Valves in the Veins

One of the classic experiments in the history of physiology was performed by William Harvey on the venous valves. This simple demonstration helped Harvey formulate his theory that blood

moves in a circular pathway around the body rather than ebbing and flowing back and forth in the vessels, as had been postulated by the physician Galen in ancient Greece.

► ACTIVITY 17.8

VENOUS VALVES

Materials

☐ 1 thick rubber band to serve as a tourniquet
☐ Venous valves

1. Tie a tourniquet around the arm above the elbow to obstruct venous return to the heart. Note the enlargement of the veins and the localized swellings that mark the position of the valves (Figure 17.13a).
2. Press one finger firmly down on a distal part of a vein. With another finger, massage the blood out of the vein toward the heart by pushing up the vein past the next valve.
3. Remove the second finger and note that the vein fills from above, but only as far as the valve (see Figure 17.13b). Press the blood from above toward the valve. What happens?
4. Remove the first finger and note the rapid filling of the vein. Starting near the elbow, push on the vein to move the blood toward the hand. What is the result?
5. Compare the filling time of the vein during rest and after exercise. ◄

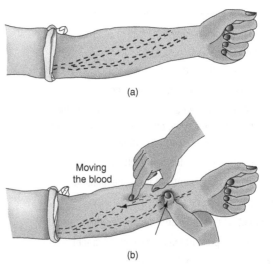

(a)

(b)

FIGURE 17.13   Harvey's experiment. (a) Location of venous valves. (b) Filling of vein only as far as valve.

Copyright © 2015, 2011, 2008 Pearson Education, Inc.

Human Cardiovascular Function

Name _____

Date _____ Section _____

Score/Grade _____

Composite Class Data

Name	Heart Rate	Blood Pressure				Cold Pressor Test	Heart Axis and Vector
		Palpatory	Auscultatory				
			Sitting	Standing	Supine		

Auscultation of Heart Sounds

1. Why are the first and second heart sounds different in intensity or pitch?

_____

_____

2. What is a heart murmur? What causes it?

_____

_____

Copyright © 2015, 2011, 2008 Pearson Education, Inc.

Measurement of Blood Pressure

1. What produces the systolic Korotkoff sound?

_____

_____

2. Why is the muffling of the sound said to indicate the diastolic pressure?

_____

_____

3. What does systolic blood pressure represent? What does diastolic blood pressure represent?

_____

_____

4. What is hypertension? What blood pressure indicates hypertension?

_____

_____

5. If your TPR = 1, what is your CO in the sitting position? (Show your calculations.)

_____

_____

6. How do you explain the changes in blood pressure that occurred when body position was altered (supine, sitting, standing)?

_____

_____

7. Explain the physiological mechanisms operating in the cold pressor test. Why would the systolic pressure rise only 10 mm Hg in a normotensive individual?

_____

_____

Arterial Pulse Wave

1. Place your recordings in the following space.

Resting   Hand raised   Hand lowered

Enter your data below:

	Delta T (time between beats) BPM (rate)	Max (height of peak)
Seated		
Arm raised		
Arm to side		
Valsalva 1		
Valsalva 2		
Valsalva 3		
Pulse with cuff 1		
Pulse with cuff 2		
Pulse with cuff 3		

Copyright © 2015, 2011, 2008 Pearson Education, Inc.

How do you explain the changes in the pulse wave when the arm is raised or lowered?

_____

_____

2. Place the pulse wave recording during the Valsalva maneuver in the following space and note the heart rate (BPM) before inhalation, just prior to exhalation, and after exhalation.

Compare the values for Delta T, BPM, and Max for Valsalva 1–3. Explain the differences observed.

_____

_____

_____

Explain the Valsalva effects on the peripheral pulse and heart rate. Why is it advisable to exhale when lifting weights?

_____

_____

3. In the following space, place your recording of the pulse during inflation and deflation of the sphyg-momanometer cuff.

Compare the values for Delta T, BPM, and Max for Pulse with cuff 1–3.

_____

_____

_____

How do you account for the pulse wave changes seen as the cuff pressure was reduced?

_____

_____

Electrocardiogram

Attach your recordings for each lead here:

Lead I   Lead II   Lead III

Copyright © 2015, 2011, 2008 Pearson Education, Inc.

Record your lead II EKG measurements for the following:

PR interval = _____ sec   PR segment = _____ sec

QT interval = _____ sec   ST segment = _____ sec

P wave amplitude = _____ mV   P wave duration = _____ sec

QRS amplitude = _____ mV   QRS duration = _____ sec

T wave amplitude = _____ mV   T wave duration = _____ sec

Attach your recording for lead II (after exercise) to the lab report.

Record your lead II EKG measurements (after exercise) for the following:

PR interval = _____ sec   PR segment = _____ sec

QT interval = _____ sec   ST segment = _____ sec

P wave amplitude = _____ mV   P wave duration = _____ sec

QRS amplitude = _____ mV   QRS duration = _____ sec

T wave amplitude = _____ mV   T wave duration = _____ Sec

Were there any differences in the measurements before and after exercise? How do you explain these differences?

_____

_____

Electrical Axis of the Heart

Draw your axis and vector in the following triangle.

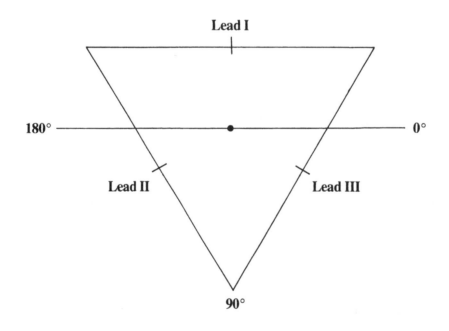

1. Mean electrical axis = _____ degrees. Heart vector length = _____ mm

2. How would the axis change during pregnancy? Why?

_____

_____

Copyright © 2015, 2011, 2008 Pearson Education, Inc.

Valves in the Veins

1. What is the importance of venous return to the pumping ability of the heart?

   _____

   _____

2. What factors aid venous return of blood to the heart?

   _____

   _____

APPLY WHAT YOU KNOW

1. How does a defibrillator help restore heart function?

   _____

   _____

   _____

2. Based on your understanding of heart and muscle function, why would it be necessary for a patient to be as still as possible while recording the ECG?

   _____

   _____

   _____

Copyright © 2015, 2011, 2008 Pearson Education, Inc.

Respiratory Function

CHAPTER 18 INCLUDES:

 3 Vernier® Activities

 2 BIOPAC® Activities

PhysioEx 9.1

For more exercises on Respiratory Function, visit PhysioEx™ (www.physioex.com) and choose Exercise 7: Respiratory System Mechanics.

OBJECTIVES

After completing this exercise, you should be able to

1. Record and explain respiratory cycles under normal and stressed conditions.
2. Measure and calculate respiratory volumes and compare data to standard measurements.

Respiratory Movements

Of the many processes occurring in our bodies each instant, those that function in the movement of oxygen to the tissues are among the most important. If tissues are deprived of oxygen for too long a time, they die; this time factor is especially critical for the cells of vital organs such as the heart and brain. Because of the importance of O_2 and CO_2, their concentration in the lungs and blood is finely regulated by a variety of receptors, reflexes, and feedback processes that control our respiratory patterns. Because the tissues involved in gas exchange are located deep within our bodies in the lungs, a tide-like movement of atmospheric air in and out of the lungs allows for both the inhalation of oxygen and the exhalation of carbon dioxide.

Both processes call for changes in the pressure of air inside the alveoli, the sac-like structures responsible for gas exchange, relative to atmospheric pressure. When the pressure in the alveoli is relatively less than atmospheric pressure, air enters the lungs; this occurs during inspiration. Conversely, when pressure in the alveoli is relatively greater than atmospheric pressure, air exits the lungs; this occurs during expiration. These changes are illustrated in Figure 18.1. The muscles involved in inspiration and expiration are illustrated in Figure 18.2. Normal, unforced inspiration calls for the contraction of the external intercostal muscles and the diaphragm, while normal, unforced expiration involves the relaxation of those muscles. Forced inspiration and expiration, typically observed during physically stressful conditions, will involve additional muscles during expiration and inspiration and will involve more complex neural control mechanisms. You can gain insight into some of these control processes by observing a person's respiratory movements and the alteration of these movements caused by various factors.

Also important in oxygen delivery is the capacity of the lungs for air intake and the ability of the lungs to move air in and out quickly. You will analyze these functions when you study the various lung volumes and capacities and conduct the pulmonary function tests.

Respiratory movements are easily recorded by using a bellows pneumograph or impedance pneumograph around the subject's chest. The experimental setups for using the impedance pneumograph are shown in Figure 18.3. The subject should be seated close to the recorder when being tested but should not look at the record. A time line should be recorded so that respiratory rates can be determined. The Vernier® system uses a respiratory belt and a gas pressure sensor to monitor respiratory movements.

Respiratory Terms

Use your textbook to become familiar with the following terms related to respiratory physiology: eupnea, polypnea, asphyxia, apnea, tachypnea, dead space, hyperpnea, anoxia, dyspnea, and hypercapnia.

Copyright © 2015, 2011, 2008 Pearson Education, Inc.

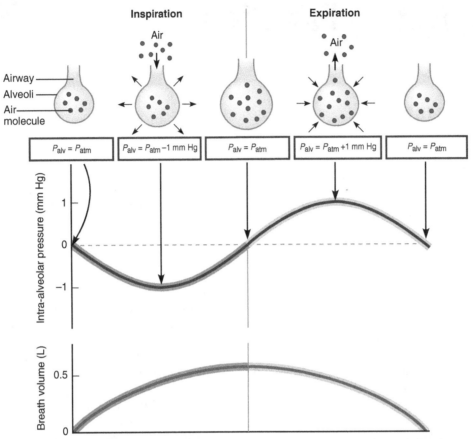

FIGURE 18.1   Changes in alveolar pressure (relative to atmospheric pressure) and breath volume during inspiration and expiration.

▶ ACTIVITY 18.1

Recording Respiratory Movements Using Impedance Pneumograph (Physiograph™ System)

Materials needed (per group)

☐ One Impedance Pneumograph

☐ One Physiograph

☐ Large paper bags

The impedance pneumograph measures the impedance between two plate electrodes applied to the thorax of the subject. A small current of 4 microamps is passed through the electrodes; the voltage across the electrodes is directly proportional to the impedance between the plates. The impedance to the passage of current is reduced when air is inhaled into the lungs and increased when air is exhaled. The resulting voltage changes are amplified and drive the pen for recording respiratory movements.

Apply electrode gel to the plate electrodes and place them on opposite sides of the thoracic cavity. Fasten them in place with a long rubber strap. Attach an additional plate electrode on the center of the chest for grounding. Connect the red and green lead wires to the recording electrodes and the black wire to the ground electrode. Connect the lead wires to the impedance pneumograph coupler by means of a shielded input extension cable.

1. Normal Respiratory Pattern

Record the subject's normal cyclic pattern of respiration for 1–2 min, first using a slow paper speed and then a faster speed. Note the amplitude of the inspiration and expiration cycles. What is the respiratory rate per minute?

2. Hyperventilation

Record normal ventilation for a few cycles at a slow paper speed. Then, at a given signal, stop the recording and have the subject breathe as fast and as deeply as possible for 30 sec. At the end of this period, obtain a record of the aftereffects

Copyright © 2015, 2011, 2008 Pearson Education, Inc.

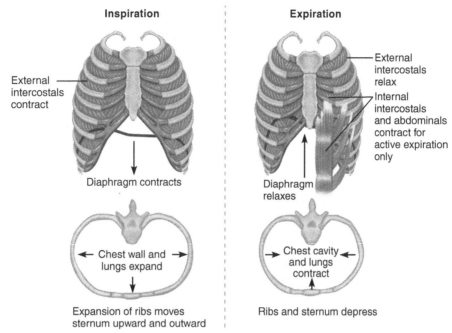

FIGURE 18.2   Respiratory muscles.

of the hyperventilation and report your observation in the Laboratory Report. The subject should allow his or her breathing to be as involuntary as possible during this posthyperventilation period. If the subject gets dizzy while hyperventilating, have him or her stop, but record the respiratory response.

What effect does hyperventilation have on the involuntary respiratory rate and amplitude of breathing? Does apnea or shallow breathing develop? What mechanism is responsible for these effects?

3. Hyperventilation in a Closed System

Repeat the hyperventilation experiment with the subject breathing in and out of a paper bag. Record the respiratory movements after hyperventilation. How does this record compare with the previous one? Is apnea as pronounced in this experiment? Explain.

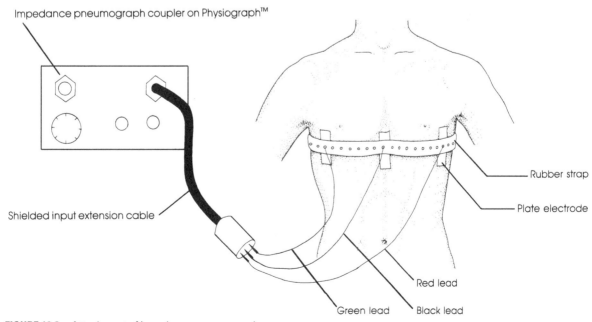

FIGURE 18.3   Attachment of impedance pneumograph.

Copyright © 2015, 2011, 2008 Pearson Education, Inc.

4. Rebreathing

Record respiratory movements while the subject breathes in and out of a paper bag for several minutes. Do not record continuously but for only 10–15 sec during each minute. The bag may be held tightly around the nose and mouth or over the entire head. Be sure to avoid leakage of air from the bag. What happens to the rate and amplitude of breathing during rebreathing of expired air? Explain the mechanism involved.

5. Effect of Mental Concentration

Record the respiratory pattern while the subject attempts to thread a needle or calculate a math problem. This experiment illustrates the effect of higher brain activity on the medullary respiratory centers. What changes in respiration can you observe?

6. Effect of Speech on Respiration

Record respiratory movements while the subject reads audibly from a book. How is the pattern changed? Record the movements while the subject reads the same paragraph silently. Is there any effect on respiration in this case?

7. Breath Holding

Obtain a short record of normal respiration; then have the subject hold his or her breath as long as possible. Record respiratory movements after the subject reaches the breaking point and resumes breathing. What differences do you notice?

8. Obstruction of Respiratory Passageways

Have the subject partially occlude his or her respiratory passageways by squeezing the nostrils. Record the respiratory patterns for a minute or so. What changes exhibit in respiratory rate and amplitude? In what disorders is obstruction of airway passages a problem?

9. Effect of Exercise

Record respiratory movements after the subject has exercised by stepping up and down on a 12-inch stool 50 times or running in place for 200 steps.[1] What factors operate during exercise to increase the rate and amplitude of breathing?

18.1 VERNIER® VERSION

Materials needed (per group)

☐ Computer

☐ Vernier® Logger Pro

☐ Vernier® respiration belt

☐ Vernier® gas pressure sensor

☐ Large paper bags

Equipment needed for this exercise will have been set up for you. Prepare the computer to measure respiratory volumes by opening the "26a Human Respiration" file in the *Biology with Computers* folder. You should see two graphs and two meter windows displayed. The upper graph will monitor changes in pressure (measured in kilopascals [kPa], a unit of pressure), whereas the lower graph will interpret this data to calculate respiratory rate in breaths per minute (bpm). The horizontal axis represents time, scaled from 0 to 180 sec.

Wrap the respiration monitor belt around the subject so that the Velcro strips are pressed together at the back. Make sure that the air bladder rests just below the sternum, at the level of the elbows. The belt should be attached to the gas pressure sensor with a rubber tube with a white Luer-lock connector. The other rubber tube has a bulb pump attached. Have the subject sit upright in a chair. Close the shut-off screw by turning it clockwise as far as you can. Fill the bladder as full as possible using the bulb pump. The belt and the bladder should press firmly against the diaphragm without being uncomfortable. Click **Collect** to test the system. As the subject breathes in and out, you should see the displayed pressure increasing and decreasing. The pressure difference should be greater than 1 kPa to register accurately. Increase the pressure in the bladder if you don't see this difference. The program is set to collect data for 180 sec; however, you can click **Stop** when necessary. The Logger Pro software also allows you to print a copy of the graph obtained using the **File Print Graph** option. Enter the name of the subject and the experimental condition in the **Comments** box to identify your graph.

---

[1]This test causes cardiovascular stress; students who have any cardiovascular difficulties should not take part unless they have permission from their physician.

Copyright © 2015, 2011, 2008 Pearson Education, Inc.

1. Normal Respiratory Pattern

Click **Collect** and record the subject's normal cyclic pattern of respiration for 3 min. Note the amplitude of the inspiration and expiration cycles. What is the respiratory rate per minute?

2. Hyperventilation

Have the subject breathe as fast and as deeply as possible for 30 sec. At the end of this period, click **Collect** and record the aftereffects of the hyperventilation for 3 min. The subject should allow his or her breathing to be as involuntary as possible during this posthyperventilation period. If the subject gets dizzy while hyperventilating, have him or her stop, but record the respiratory response.

What effect does hyperventilation have on the involuntary respiratory rate and amplitude of breathing? Does apnea or shallow breathing develop? What mechanism is responsible for these effects?

3. Hyperventilation in a Closed System

Repeat the hyperventilation experiment with the subject breathing in and out of a paper bag. Click **Collect** to record the respiratory movements after hyperventilation. How does this record compare with the previous one? Is apnea as pronounced in this experiment? Explain.

4. Rebreathing

Click **Collect** to record respiratory movements while the subject breathes in and out of a paper bag for several minutes. Do not record continuously but for only 10 to 15 sec during each minute. The bag may be held tightly around the nose and mouth or over the entire head. Be sure to avoid leakage of air from the bag. What happens to the rate and amplitude of breathing during the rebreathing of expired air? Explain the mechanism involved.

5. Effect of Mental Concentration

Click **Collect** and record the respiratory pattern while the subject attempts to thread a needle or calculate a math problem. This experiment illustrates the effect of higher brain activity on the medullary respiratory centers. What changes in respiration are observed?

6. Effect of Speech on Respiration

Click **Collect** and record respiratory movements while the subject reads audibly from a book. How is the pattern changed? Record the movements while the subject reads the same paragraph silently. Is there any effect on respiration in this case?

7. Breath Holding

Click **Collect** and obtain a short record of normal respiration; then have the subject hold his or her breath as long as possible. Continue recording respiratory movements after the subject reaches the breaking point and resumes breathing. What differences do you notice?

8. Obstruction of Respiratory Passageways

Have the subject partially occlude his or her respiratory passageways by squeezing the nostrils. Click **Collect** and record the respiratory patterns for a minute or so. What changes do you see in respiratory rate and amplitude? In what disorders is obstruction of airway passages a problem?

9. Effect of Exercise

Click **Collect** to record respiratory movements after the subject has exercised by stepping up and down on a 12-inch stool 50 times or running in place for 200 steps.[2] What factors operate during exercise to increase the rate and amplitude of breathing? ◄

CLINICAL APPLICATION

Respiratory Movement and Obstructive Pulmonary Disorders

The upper respiratory tract and the bronchial tree are the passageways to the lungs. Normally, the body experiences little resistance to the movement of air into the alveoli, where gas exchange takes place. Occasionally, these passageways can obstruct the movement of air to and from the lungs, making ventilation inefficient and reducing gas exchange. **Obstructive pulmonary disorder** is a term that describes a number of conditions that reduce ventilation capacity. Emphysema and chronic bronchitis are conditions generally known as COPD,

---

[2]This test causes cardiovascular stress; students who have any cardiovascular difficulties should not take part unless they have permission from their physician.

Copyright © 2015, 2011, 2008 Pearson Education, Inc.

chronic obstructive pulmonary disorders. Patients with emphysema experience reduced lung ventilation caused by a loss of elasticity and an increase in compliance of the lungs. Chronic bronchitis is a persistent inflammation of the bronchi, which thickens the airway lining and increases resistance to the flow of air in and out of the lungs. Asthma is an acute condition that occurs when a patient inhales allergens and the smooth muscle lining of the respiratory passage contracts in response. Physicians might prescribe various bronchodilators to alleviate the symptoms of these obstructive disorders.

Respiratory Volumes

The total capacity of the lungs is divided into various volumes and capacities according to their function in the intake or exhalation of air. For a proper understanding of respiratory processes, you must become familiar with the following volumes and capacities. The values provided below are those for a 70-kg male. The values for females are usually about 20–25% less. Considerable variation, correlated with age, sex, height, and weight, is usual.

As shown in Figure 18.4, the total amount of air one's lungs can hold can be subdivided into four **volumes**, defined as follows:

Tidal volume (TV). The amount of air inspired or expired during normal, quiet respiration. This volume is typically 500 ml for both males and females.

Inspiratory reserve volume (IRV). The amount of air that can be forcefully inspired beyond that taken in during abnormal inspiration. This volume is typically 3000 ml for a male and 1900 ml for a female.

Expiratory reserve volume (ERV). The amount of air that can be forcefully expired following a normal expiration. This volume is typically 1100 ml for a male and 700 ml for a female.

Residual volume (RV). The amount of air that remains trapped in the lungs after a maximal expiratory effort. This volume cannot be measured using the spirometer but is assumed to be, on the average, about 1200 ml for a male and 1100 ml for a female. It can range between 1 and 2.5 L.

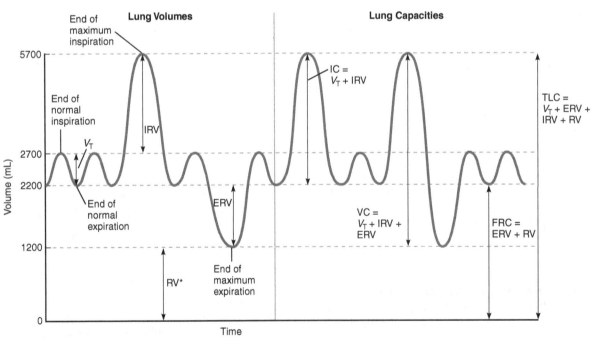

Normal lung volumes and capacities for a healthy 70-kg male

Lung Volumes	Lung Capacities
V_T = Tidal volume = 500 mL	IC = Inspiratory capacity = V_T + IRV = 3500 mL
IRV = Inspiratory reserve volume = 3000 mL	VC = Vital capacity = V_T + IRV + ERV = 4500 mL
ERV = Expiratory reserve volume = 1000 mL	FRC = Functional residual capacity = ERV + RV = 2200 mL
RV = Residual volume* = 1200 mL	TLC = Total lung capacity = V_T + ERV + IRV + RV = 5700 mL

*Cannot be measured by spirometry

FIGURE 18.4   Lung volumes and capacities measured using spirometry.

Copyright © 2015, 2011, 2008 Pearson Education, Inc.

In addition to the four volumes, which do not overlap, there are four **capacities**, which are combinations of two or more volumes.

Total lung capacity (TLC). The total amount of air the lungs can contain—the sum of all four volumes.

Vital capacity (VC). The maximal amount of air that can be forcefully expired after a maximal inspiration.

Functional residual capacity (FRC). The amount of air remaining in the lungs after a normal expiration.

Inspiratory capacity (IC). The maximal amount of air that can be inspired after a normal expiration. The milliliter values given for these volumes and capacities in Figure 18.4 are for a normal adult male. In the female, they are all 20–25% smaller.

The respiratory volumes can be measured with a simple instrument called a **spirometer**. This consists of a lightweight metal bell inverted in a drum filled with water. A mouthpiece and hose allow the collection of air in the inverted bell. In these experiments, you use your own disposable mouthpiece and a bacterial filter. Record your results in the table in the Laboratory Report. In addition to this traditional method, the **Vernier spirometer** and the **BIOPAC spirometer** systems (Figure 18.5) use a pneumotachometer to measure slight differences in pressure as air is blown through a fine mesh. The small pressure difference is proportional to flow rate. Two small plastic tubes on either side of the mesh transmit this pressure difference to a transducer that converts the pressure signal into a changing voltage

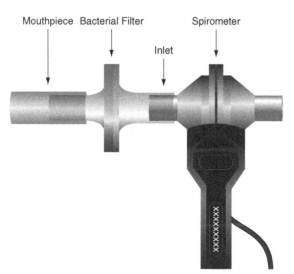

FIGURE 18.5   A spirometer with a bacterial filter and mouthpiece, used in the BIOPAC® and Vernier® systems.

that is recorded by a data acquisition system. Computer software will interpret and display these signals on a screen.

⚠ CAUTION!

Do not perform this experiment if you have any respiratory ailment. Make sure that you use a new mouthpiece and bacterial filter and that these are appropriately disposed of after use.

► ACTIVITY 18.2

Measuring Respiratory Volume Using the Spirometer

Refer to Figure 18.4 to understand what you are measuring. You may need to take measurements from different cycles and obtain an average if there are significant differences between breathing cycles.

1. Tidal Volume (TV)

Set the spirometer dial to zero. After a normal inspiration, place your mouth over the mouthpiece and exhale normally into the spirometer. You will have to make a conscious effort not to exceed your normal volume. Read the amount exhaled on the dial. Multiply your tidal volume by your respiratory rate per minute to give your resting **respiratory minute volume**. Have your lab partner count your respiratory cycles for 1 min while you are seated at rest. **Enter these values in the table in the Laboratory Report.**

2. Expiratory Reserve Volume (ERV)

Set the spirometer dial to zero. After a normal expiration, place your mouth over the mouthpiece and forcefully exhale as much air as possible into the spirometer. **Enter these values in the table in the Laboratory Report.**

3. Vital Capacity (VC)

Set the spirometer dial to zero. Inhale as deeply as possible; then place your mouth over the mouthpiece, hold your nose, and exhale into the spirometer with a maximal effort. Repeat the measurement three times and record the largest volume. Using nomograms 1 and 2 in Appendix A, determine your **predicted vital capacity (VC)** based on your age, height, and sex. How does your predicted VC compare with your measured VC? **Enter these values in the table in the Laboratory Report.**

4. Inspiratory Reserve Volume (IRV) and Inspiratory Capacity (IC)

From the three previous volume measurements you can now calculate the IRV and the IC. **Enter these values in the table in the Laboratory Report.**

Copyright © 2015, 2011, 2008 Pearson Education, Inc.

5. Heymer Test of Respiratory Reserve

Take five deep breaths and then hold your breath as long as possible after the last inspiration. The breath-holding time gives an indication of your functional respiratory reserve and the efficiency of your respiratory system. Normal values are 50–70 sec for men and 50–60 sec for women. This test should be performed two or three times and the average taken. The Heymer test is often a better index of respiratory reserve than the traditional vital capacity measurement. **Enter these values in the table in the Laboratory Report.**

The principal value of these pulmonary measurements lies in following volume changes caused by either disease or recovery from a disease. For example, the VC is found to decrease in left heart disease. The decrease is due to blood congestion in the lung capillaries, which in turn leads to pulmonary edema and a decrease in VC. As the person recovers, his or her heart becomes stronger, pulmonary congestion and edema decrease, and the VC increases. The VC also decreases in paralytic polio, owing to partial paralysis of respiratory muscles, and in various other respiratory diseases.

18.2 VERNIER® VERSION

Materials needed (per group)

- [ ] Computer
- [ ] Vernier® Logger Pro®
- [ ] Vernier® spirometer
- [ ] Disposable mouthpiece
- [ ] Disposable bacterial filter
- [ ] Nose clip

Equipment needed for this exercise will have been set up for you. Prepare the computer to measure respiratory volumes by opening the "19 Lung Volumes" file in the "Human Physiology with Vernier" folder.

Zero the sensor by holding the spirometer upright and very still and then clicking **Zero**. The spirometer is now ready.

Data Collection

a. Put on the nose clip.
b. Breathe in normally, click **Collect**, place your mouth over the mouthpiece, and exhale normally into the spirometer. Continue breathing in and out as normally as possible for six cycles. You will have to make a conscious effort not to exceed your normal volumes. After these cycles, inhale as deeply as possible (maximal inspiration) and then exhale as fully as possible (maximal exhalation). Follow this with at least two normal breathing cycles. Click **Stop**. Your record should resemble that seen in Figure 18.4.
c. Have your lab partner count your respiratory cycles for 1 min while you are seated at rest.

Data Analysis

Refer to Figure 18.4 to understand what you are measuring. You might need to take measurements from different cycles and obtain an average if there are significant differences between breathing cycles.

1. Tidal Volume (TV)

Choose a peak and valley from the first six cycles. Point your cursor to a peak, click, drag, and release the button at the bottom of the valley. At the bottom right of the screen, you should see a Δy value displayed. This is your tidal volume. Multiply your tidal volume by your respiration rate per minute to give you your resting **respiratory minute volume**. **Enter these values in the table in the Laboratory Report.**

2. Expiratory Reserve Volume (ERV)

Move the cursor to the valley that represents your maximal expiration. Click and drag up the side of the peak until you reach the level of the valleys graphed during normal breathing. The Δy value displayed at the bottom right of your graph is the expiratory reserve volume. **Enter these values in the table in the Laboratory Report.**

3. Vital Capacity (VC)

Move the cursor to the peak that represents your maximal inspiration. Click and drag down the side of the peak until you reach the level of the valley associated with your maximal expiration. The Δy value displayed at the bottom right of your graph is your vital capacity (VC).

Using nomograms 1 and 2 in Appendix A, determine your **predicted vital capacity** based on your age, height, and sex. How does your predicted VC compare with your measured VC? **Enter these values in the table in the Laboratory Report.**

4. Inspiratory Reserve Volume (IRV)

Move the cursor to the peak that represents your maximal inspiration. Click and drag down the side of the peak until you reach the level of the peaks associated with your normal breathing. The Δy value displayed at the bottom right of your graph is your inspiratory reserve volume. **Enter these values in the table in the Laboratory Report.**

Copyright © 2015, 2011, 2008 Pearson Education, Inc.

5. Inspiratory Capacity (IC)

Move the cursor to the peak that represents your maximal inspiration. Click and drag down the side of the peak until you reach the level of the valleys associated with your normal breathing. The Δy value displayed at the bottom right of your graph is your inspiratory capacity. **Enter these values in the table in the Laboratory Report.**

6. Heymer Test of Respiratory Reserve

Take five deep breaths and then hold your breath as long as possible after the last inspiration. The breath-holding time gives an indication of your functional respiratory reserve and the efficiency of your respiratory system. Normal values are 50–70 sec for men and 50–60 sec for women. This test should be performed two or three times and the average taken. The Heymer test is often a better index of respiratory reserve than the traditional vital capacity measurement. **Enter these values in the table in the Laboratory Report.**

The principal value of these pulmonary measurements lies in following volume changes caused by either disease or recovery from a disease. For example, the VC is found to decrease in left heart disease. The decrease is due to blood congestion in the lung capillaries, which in turn leads to pulmonary edema and a decrease in VC. As the person recovers, his or her heart becomes stronger, pulmonary congestion and edema decrease, and the VC increases. The VC also decreases in paralytic polio, owing to partial paralysis of respiratory muscles, and in various other respiratory diseases.

18.2 BIOPAC® VERSION

Materials needed (per group)

☐ BIOPAC® airflow transducer (SS11LA)
☐ Disposable mouthpieces
☐ Bacteriological filters
☐ Nose clip
☐ Calibration syringe
☐ Computer
☐ BIOPAC® Student Lab 3.7 or higher
☐ BIOPAC® data acquisition system

Equipment needed for this exercise will have been set up for you. Because the airflow transducer is sensitive to gravity, make sure that it is held upright while calibrating and recording data.

Calibrating the Airflow Transducer

1. Turn the computer on first and then turn on the BIOPAC® data acquisition system.

2. Place a filter onto the end of the calibration syringe.
3. Insert the calibration syringe/filter assembly into the inlet of the airflow transducer.
4. Start the BIOPAC® Student Lab program and choose lesson 12 (L12-Lung-1).
5. Type in your file name, using your first name and last name initial to identify this file uniquely, and click **OK**.
6. Pull the calibration syringe plunger all the way out and hold the assembly upright. Make sure you use one hand to move the plunger and the other hand to grip the barrel of the syringe. Do not hold onto the airflow transducer; it should hang freely off the syringe.
7. There are two stages to the calibration. Once you click **calibrate**, stage 1 will run independently for 8 sec and end with an alert asking whether you have read the instruction in the journal below. After you have read and understood the directions, click **Yes**. This will start the second stage of the calibration.
8. Push the cylinder in completely, wait 2 sec, and pull the cylinder out. Repeat this procedure four more times and then click **End Calibration**. Your screen should resemble that seen in Figure 18.6. Redo the calibration if necessary.

Data Collection

Put on the nose clip and use a new mouthpiece and filter for each subject.

1. Click **Record**.
2. Breathe normally for five breaths (inhale–exhale cycle = 1 breath) and then inhale as deeply as you can.
3. Then exhale as deeply as you can.
4. Breathe normally for five more breaths.
5. Click **Stop**.

If your data resembles the upper panel in Figure 18.7, click **Done**; otherwise, click **Redo**. When you click **Done**, your data is automatically saved under the file name you used in step 5 of the calibration procedure.

A pop-up window will ask if you need to "record from another subject." Select "yes" to avoid recalibrating the air flow transducer. Use a unique file name and repeat the preceding steps 1–5 for each additional subject. When all recordings are completed, proceed to data analysis.

Data Analysis

Refer to Figure 18.4 to understand what you are measuring. You might need to take measurements from different cycles and obtain an average if there are significant differences between breathing cycles.

Copyright © 2015, 2011, 2008 Pearson Education, Inc.

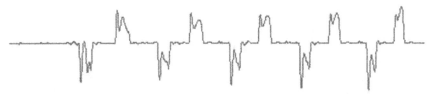

FIGURE 18.6 Calibration screen for BIOPAC® airflow sensor.

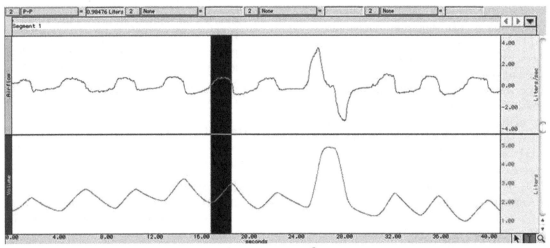

FIGURE 18.7 Lung volume recording and analysis screen with BIOPAC®.

1. Enter the **Review Saved Data** mode and select the file you intend to analyze.
2. You will make all your measurements on the Volume-channel; if two channels appear on-screen, you can hide the upper Airflow channel by using option + click (Mac) or ctrl + click (Windows) over the channel number box (Figure 18.7).
3. Set up the measurement boxes on the top of the screen so that these options are shown:

   CH 2 P-P (max value in selected area minus minimum value in that area)

   CH2 Max Maximum value

   CH 2 Min Minimum value

   CH 2 Delta Difference between first and last point in selected area

Refer to Figure 18.4 to understand what you are measuring. You might need to take measurements from different cycles and obtain an average if there are significant differences between breathing cycles.

4. To find the tidal volume (TV), use the I-beam cursor to choose a peak and valley from the first five cycles (lower panel, Figure 18.7). Point your cursor to a peak; click, drag, and release the button at the bottom of the valley. The **CH 2 P-P** is your tidal volume. Multiply your tidal volume by your respiration rate per minute to yield your resting respiratory minute volume. Enter these values in the table in the Laboratory Report.
5. To find the expiratory reserve volume (ERV), move the I-beam cursor to the lowest-point

Copyright © 2015, 2011, 2008 Pearson Education, Inc.

valley that represents your maximal expiration. Click and drag up the side of the peak until you reach the level of the next valley graphed during normal breathing. The **CH 2 Delta** value displayed is the expiratory reserve volume. Enter these values in the table in the Laboratory Report.

6. To find the vital capacity (VC), move the I-beam cursor to the peak that represents your maximal inspiration. Click and drag down the side of the peak until you reach the lowest level of the valley associated with your maximal expiration. The **CH 2 Delta** value displayed at the bottom right of your graph is your vital capacity (VC). Enter these values in the table in the Laboratory Report.

   Using nomograms 1 and 2 in Appendix A, determine your **predicted vital capacity** based on your age, height, and sex. How does your predicted VC compare with your measured VC?

7. To find the inspiratory reserve volume (IRV), move the I-beam cursor to the peak that represents your maximal inspiration. Click and drag down the side of the peak until you reach the level of the peaks associated with your normal breathing. The **CH 2 Delta** value displayed is your inspiratory reserve volume. **Enter these values in the table in the Laboratory Report.**

8. To find the inspiratory capacity (IC), move the cursor to the peak that represents your maximal inspiration. Click and drag down the side of the peak until you reach the level of the valleys associated with your normal breathing. The **CH 2 Delta** value displayed is your inspiratory capacity. **Enter these values in the table in the Laboratory Report.**

9. For the Heymer test of respiratory reserve, take five deep breaths and then hold your breath as long as possible after the last inspiration. The breath-holding time gives an indication of your functional respiratory reserve and the efficiency of your respiratory system. Normal values are 50–70 sec for men and 50–60 sec for women. This test should be performed two or three times and the average taken. The Heymer test is often a better index of respiratory reserve than the traditional vital capacity measurement. **Enter these values in the table in the Laboratory Report.**

The principal value of these pulmonary measurements lies in following volume changes caused by either disease or recovery from a disease. For example, the VC is found to decrease in left heart disease. The decrease is due to blood congestion in the lung capillaries, which in turn leads to pulmonary edema and a decrease in VC. As the person recovers, his heart becomes stronger, pulmonary congestion and edema decrease, and the VC increases. The VC also decreases in paralytic polio, owing to partial paralysis of respiratory muscles, and in various other respiratory diseases. ◄

Pulmonary Function Tests

In most cases, measurements of pulmonary volumes are of limited value because they are simply measurements of the anatomical size of lung compartments. They tell virtually nothing about the ability to move air in and out of the lungs, a critical factor in the delivery of oxygen to the blood. Consequently, in recent years, we have seen increased use of **pulmonary function tests** as indexes of respiratory efficiency. Such tests are valuable because they give some measure of (1) lung compliance, or elasticity; (2) airway resistance; and (3) respiratory muscle strength. These three factors determine how much air a person can move into the lungs per unit of time, and this is what the pulmonary function tests measure.

► ACTIVITY 18.3
Measuring Pulmonary Function

☐ 1 Douglas bag
☐ Alcohol swabs
☐ 1 Flowmeter
☐ 1 timed vitalometer or recording vitalometer

1. Maximal Voluntary Ventilation (MVV) or Maximal Breathing Capacity (MBC)
This is a measurement of the maximal volume of air that can be moved through the lungs in 1 min. Clamp your nose, breathe as rapidly and deeply as possible for 15 sec through a low-resistance respiratory valve, and collect the expired air in a Douglas bag. The volume of air forced out in 15 sec is determined by connecting the Douglas bag to a flow meter and squeezing the bag to push all of the air collected through the flow meter (Figure 18.8). Multiply the volume of air collected in 15 sec by four to convert to liters (L)

Copyright © 2015, 2011, 2008 Pearson Education, Inc.

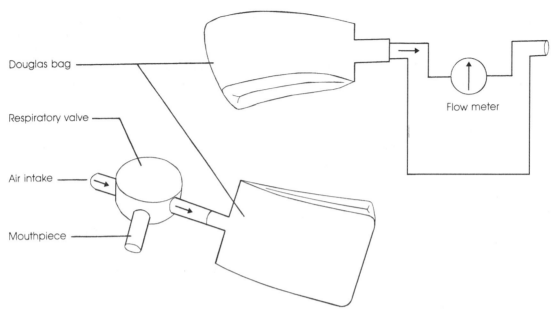

FIGURE 18.8   Douglas bag and flow meter. Arrows denote air movement.

per minute. For college-aged men, the normal MVV is 140–180 L/min; for college-aged women, it is 80–120 L/min. Tables and formulas based on age and body surface area are available for calculating a person's predicted MBC. Two of the commonly used formulas are as follows:

$$\text{Males:}\quad \text{MVV} = [86.5 - (0.522 \times \text{Age})] \\ \times \text{Body surface area (m}^3\text{)}$$
$$\text{Females:}\ \text{MVV} = [71.3 - (0.474 \times \text{Age})] \\ \times \text{Body surface area (m}^2\text{)}$$

The MVV is a good test of the ability to move air rapidly but has the disadvantage of requiring a strenuous effort, so that considerable coaching and motivation are needed to obtain valid results. In addition, many feeble or ill persons are unable to exert themselves for as long as 15 sec. Because of these disadvantages, several single-breath tests have been devised that give almost the same information as the MVV test.

2. Forced Expiratory Volume (FEV) or Timed Vital Capacity (TVC)

This test measures the volume of air expired in 1, 2, or 3 sec during a maximal exertion. This volume is then changed to percent of total vital capacity expired during the entire expiratory period. A normal person should be able to exhale 83% of his or her VC forcefully in 1 sec ($\text{FEV}_{1.0} = 83\%$), 94% of the VC in 2 sec, and 97% of the VC in 3 sec. The 1-sec forced expiratory volume ($\text{FEV}_{1.0}$) has been found to correlate best with MVV measurements. The correlation coefficients between them are 0.88–0.92. Typically, the MVV is about 30–40 times the value of $\text{FEV}_{1.0}$.

You can also measure the FEV with a **timed vitalometer** or a **recording vitalometer**. Simply take a deep breath and then exhale as forcefully and fully as possible into the vitalometer. The timed vitalometer records, on a dial, the total VC and the volume of air forced out in 0.5, 0.75, 1, 2, or 3 sec. The recording vitalometer makes a graph of the forced exhalation, and the FEV can then be calculated from the graph. As in the other sets, the predicted FEV may be calculated from equations and tables based on age and height (nomograms 1 and 2 in Appendix A). You can also measure the FEV using the Vernier® or BIOPAC® air flow transducers, as described below.

18.3 VERNIER® VERSION (FOR MEASURING FORCED EXPIRATORY VOLUME ONLY)

Prepare the computer to measure respiratory volumes by opening the "21 Analyze Lung Function" file in the "Human Physiology with Vernier" folder. Once again, make sure that you use a new mouthpiece and bacterial filter for this experiment.

Zero the sensor by holding the spirometer upright and very still and then click **Zero**. The spirometer is now ready.

a. Put on the nose clip.
b. Click **Collect**.
c. Inhale as deeply as possible and then exhale as quickly and fully as you can into the spirometer. Without removing your mouth from the mouthpiece, inhale as deeply as possible. Your exhalation will trigger data collection, and collection will be terminated automatically. The flow volume loop graph

Copyright © 2015, 2011, 2008 Pearson Education, Inc.

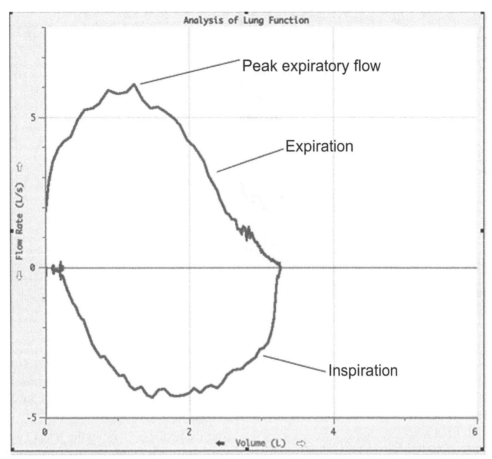

FIGURE 18.9   Flow volume loop graph.

Source: © Vernier Software & Technology. Used with permission.

obtained should resemble that in Figure 18.9. Additionally, numerical data are collected and displayed in the graph to the left of the screen.

d. Scroll down the data table and record the volumes associated with 1.00, 2.00, and 3.00 sec.

e. Click the **Examine** button. Move the examine line to the rightmost point of the flow. Record the volume value to the nearest 0.01 L.

f. As in the other sets, the predicted FEV may be calculated from equations and tables based on age and height (nomograms 1 and 2 in Appendix A). Is your predicted FEV close to your measured FEV?

18.3 BIOPAC® VERSION (FOR MEASURING FORCED EXPIRATORY VOLUME [FEV] AND MAXIMUM VOLUNTARY VENTILATION [MVV] ONLY)

Materials needed (per group)

☐ BIOPAC® airflow transducer (SS11LA)

☐ Disposable mouthpieces

☐ Bacteriological filters

☐ Nose clip

☐ Calibration syringe

☐ Computer

☐ BIOPAC® Student Lab 3.7.5 or higher

☐ BIOPAC® Data Acquisition system

Equipment needed for this exercise will have been set up for you. Because the airflow transducer is sensitive to gravity, make sure that it is held upright while calibrating and recording data.

1. Start the BIOPAC® Student Lab program and choose Lesson 13 (L13-Lung-2).

2. Type in your file name, using your first name and last name initial to identify this file uniquely, and click **OK**.

If you are not continuing directly from Activity 18.2, calibrate the BIOPAC® airflow transducer as described in Activity 18.2, steps 6–8. The graph should resemble Figure 18.6.

Data Collection (FEV)

Put on the nose clip and use a new mouthpiece and filter for each subject.

1. Begin breathing through the airflow transducer.

Copyright © 2015, 2011, 2008 Pearson Education, Inc.

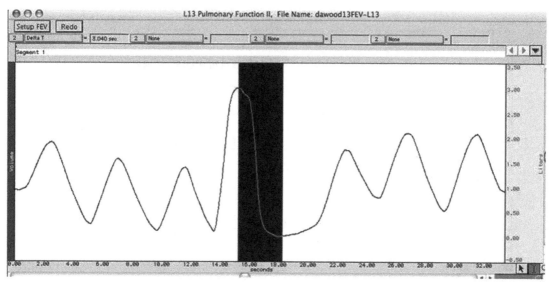

FIGURE 18.10   FEV recording screen with BIOPAC®.

2. Click on **Calculate FEV** (older software will say Record FEV).
3. Perform these procedures:
   a. Breathe normally for three cycles.
   b. Inhale as deeply as you can and hold your breath for 1 sec.
   c. Exhale as completely as you can and then breathe normally for three more cycles.
   d. Click **Stop**.
   e. Click **Redo** if necessary.
4. Use the I-beam cursor to select the area of maximum exhale. Start at the instance of exhalation and select an area at least 3 sec in length. Use the **ΔT (Delta Time)** at the top of the panel to determine the amount of time selected.
5. Click **Continue** (older software will not have this button).
6. Click **Record MVV** (older software will say Begin MVV). The data on the screen should resemble Figure 18.10.
7. Click **Done** unless you need to redo your data collection.

Your data will be saved in the "Data Files" folder when you click **Done**. An FEV extension will be added after the file name.

Data Collection (MVV)

If continuing with the measurement of MVV, click **Begin MVV**.

Place a nose clip on your nose and begin breathing through the transducer mouthpiece.

1. Click **Record MVV**.
2. Breather normally for five cycles and then quickly and deeply for about 12–15 sec.

Caution: This may cause individuals to feel lightheaded or dizzy. Stop if this feeling prevents you from wanting to continue.

3. Breathe normally again for five more cycles and then click **Stop**.

Your data will be saved in the "Data Files" folder when you click **Done**. An MVV extension will be added after the file name.

Data Analysis (FEV)

1. Enter the **Review saved data** mode and choose the correct file (with an FEV extension).
2. To turn grids on, click the **File** menu, select **Display Preferences**, choose **Grids**, select **Show grids**, and then click **OK**.
3. The first two measurement boxes at the top of the panel are:

   CH 2 Delta T
   CH 2 p-p

4. Use the I-beam cursor to select the area from time zero to the end of the recording (Figure 18.11). The **p-p** value represents your vital capacity (VC). **Enter these values in the table in the Laboratory Report.**
5. Use the I-beam cursor to select the area for the first 1-sec interval. The **p-p** value can be used to calculate FEV_1, the percentage of VC expired in the first second of exhalation.
6. Similarly, use the I-beam cursor to select the first 2-sec and then the first 3-sec intervals. Calculate FEV_2 and FEV_3 accordingly. **Enter these values in the table in the Laboratory Report.**

Copyright © 2015, 2011, 2008 Pearson Education, Inc.

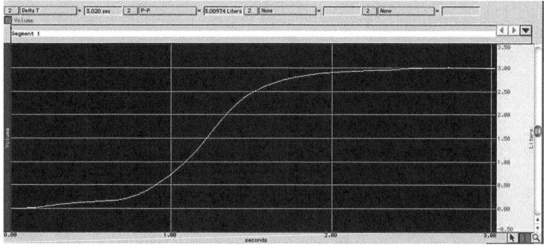

FIGURE 18.11   FEV analysis screen with BIOPAC®.

7. If other data files need to be analyzed, repeat steps 1–5 in the **Data Analysis** section. As in the other sets, the predicted FEV may be calculated from equations and tables based on age and height (nomograms 1 and 2 in Appendix A). Is your predicted FEV close to your measured FEV?

Data Analysis (MVV)

1. Enter the **Review saved data** mode and choose the correct file (MVV extension).
2. The measurement boxes should read:

   CH 2 Delta T
   CH 2 p-p

3. Use the I-beam cursor to select a 12-sec area from the rapid-breathing recording. You should be able to count the number of cycles in this area. Your screen should resemble Figure 18.12.

4. Place a marker at the end of the selected area by clicking in the marker area just above the data panel, using the marker tool (the arrow adjacent to the I-beam tool).
5. Use the I-beam tool to select each complete individual (Figure 18.13) cycle from the selected area and, one cycle at a time, record the **p-p value** for each cycle. **Record these values in the Laboratory Report.**
6. Calculate the MVV for each individual using the steps in the Laboratory Report. ◄

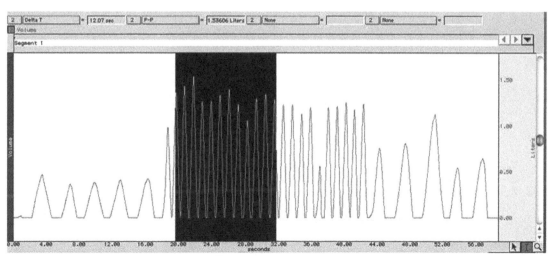

FIGURE 18.12   MVV analysis screen with BIOPAC®.

Copyright © 2015, 2011, 2008 Pearson Education, Inc.

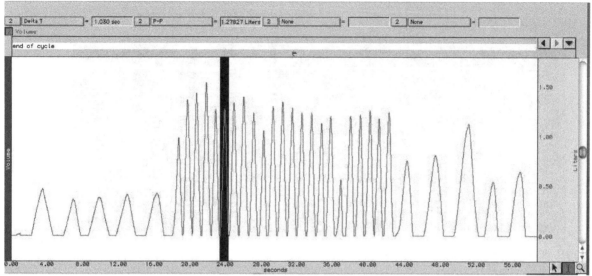

FIGURE 18.13 MVV analysis screen with one cycle selected (BIOPAC®).

Copyright © 2015, 2011, 2008 Pearson Education, Inc.

Respiratory Function

Name _____

Date _____ Section _____

Score/Grade _____

Respiratory Movements

1. What happens to the respiratory rhythm following hyperventilation? _____
_____
_____

2. What causes the apnea that sometimes occurs after hyperventilation? _____
_____
_____

3. Why does a person often get lightheaded and dizzy after blowing up a balloon or blowing a horn vigorously? What mechanism causes these sensations? _____
_____
_____

4. Explain the difference in a person's respiratory movements when hyperventilation takes place in and out of a paper bag. _____
_____
_____

5. How does reading or concentration alter respiratory movements? What is the teleological explanation for such alterations? _____
_____
_____

6. Which of the following has the most profound effect on respiration: oxygen, hydrogen ions, or carbon dioxide? Which receptors are most sensitive to changes in blood chemistry? _____
_____
_____

7. Compare the changes in blood pH, carbon dioxide, and oxygen during sustained exercise. What causes the large increase in respiratory minute volume during exercise? _____
_____
_____
_____
_____

Respiratory Volumes

Use Figure 18.4 to help you calculate IRV and IC in the following chart.

| Name | TV | Resp. Min. Vol. | ERV | VC | Predicted VC | Calculated | | Heymer Test | MVV | Predicted MVV | $FEV_{1.0}$ | Predicted $FEV_{1.0}$ |
						IRV	IC					

Copyright © 2015, 2011, 2008 Pearson Education, Inc.

| Name | TV | Resp. Min. Vol. | ERV | VC | Predicted VC | Calculated | | Heymer Test | MVV | Predicted MVV | FEV$_{1.0}$ | Predicted FEV$_{1.0}$ |
						IRV	IC					
Class Average												

MVV Calculations (BIOPAC®)

Cycle	CH 2 (p-p) (liters)
1	
2	
3	
4	
5	
6	
7	
8	
9	
10	
11	
12	
13	
14	

Copyright © 2015, 2011, 2008 Pearson Education, Inc.

1. Number of cycles in 12-sec interval (A): _____

2. Number of cycles per minute (RR): A × 5 = _____

3. Average volume per cycle (AVPC) = _____

4. MVV = AVPC × RR = _____

5. Vital capacity can be used to evaluate recovery from a myocardial infarction (MI). Explain why the measurement of vital capacity can be used as an index of recovery from a heart attack. _____
   _____
   _____

6. Which respiratory disorders produce an increase in dead space volume? _____
   _____
   _____

   How would this increase in dead space affect cellular respiration and metabolism? _____
   _____
   _____

7. Why is there an age-related decline in respiratory function?
   _____
   _____

8. How and why would an asthmatic attack have an impact on the FEV and MVV? _____
   _____
   _____

9. What drugs are used to counteract an asthmatic attack, and which part of the autonomic nervous system do they mimic in function?
   _____
   _____

APPLY WHAT YOU KNOW

Use your textbook to explain the pathophysiology of the following respiratory disorders:

Pneumonia: _____
_____
_____

Asthma: _____
_____
_____

Atelectasis: _____
_____
_____

Pleurisy: _____
_____
_____

Emphysema: _____
_____
_____

Copyright © 2015, 2011, 2008 Pearson Education, Inc.

Blood Physiology I: Erythrocyte Functions

19

CHAPTER 19 INCLUDES:

 Vernier 1 Vernier® Activity

PhysioEx 9.1 For more exercises on Blood Physiology, visit PhysioEx™ (www.physioex.com) and choose Exercise 11: Blood Analysis.

OBJECTIVES

After completing this exercise, you should be able to

1. Determine the hematocrit and hemoglobin concentration of a blood sample.
2. Use a hemocytometer to determine a red blood cell count.
3. Calculate oxygen carrying capacity, mean corpuscular hemoglobin concentration, and the mean corpuscular volume of a sample of blood.
4. Examine the effects of vasoactive agents on a capillary network in the frog.

⚠ CAUTION!

Parts of this activity might involve working with human blood. You should handle only your own blood. Dispose of all supplies (cotton, gauze, lancets, and so on) that are exposed to blood in properly marked containers. ALL BODY FLUIDS AND SUPPLIES MUST BE TREATED AS POTENTIALLY INFECTIOUS. Read the precautions for handling blood inside the front cover of this text.

Functions of Blood

Blood serves the cells of complex organisms in the same way that the aquatic environment serves unicellular organisms. That is, it provides a medium for the maintenance of homeostasis in the cells' environment. To do this in complex organisms, blood must function as a transportation system, bringing nutrients and oxygen to the cells and removing waste and carbon dioxide from the interstitial fluid around the cells. This transportation system also links the various organs of the body together, integrating them through the action of hormones. Blood also performs other functions that are not as obvious such as providing buffers for acid–base balance, destroying foreign organisms through phagocytosis and antibody action, distributing and conserving body heat, and preventing its own loss through hemostatic (coagulation) mechanisms.

Because hemoglobin is the significant molecule involved in the transportation of oxygen to the tissues, it is important to understand the factors affecting the affinity of the hemoglobin molecule for oxygen. Because air is composed of a number of separate gases, the total atmospheric pressure is determined by summing up the individual pressures of each of the component gases. Consequently, the partial pressure of an individual gas is the fractional concentration of that gas times the total atmospheric pressure (760 mm Hg) exerted by the gas mixture. For example, because oxygen is typically 21% of the atmospheric gases and carbon dioxide is 0.03%, the partial pressure of oxygen (PO_2) is:

$$PO_2 = 0.21 \times 760 \text{ mm Hg} = 160 \text{ mm Hg}$$

and the partial pressure of CO_2 (PCO_2) is:

$$PCO_2 = 0.0003 \times 760 \text{ mm Hg} = 0.23 \text{ mm Hg}$$

The ability of hemoglobin to either load or unload oxygen is usually described by the hemoglobin-oxygen dissociation curve shown in Figure 19.1. A leftward shift of the curve indicates an increased affinity for oxygen, while a rightward shift indicates a decreased affinity for oxygen. Temperature, pH, and PO_2, and PCO_2 levels all affect the ability of hemoglobin to transport oxygen, PO_2 and PCO_2.

🔴 STOP AND THINK

Based on your knowledge of conditions existing in the alveoli of the lungs and in metabolically active tissues such as skeletal muscle, would you expect a rightward or leftward shift of the dissociation curve in each of those locations? Make sure you completely

Copyright © 2015, 2011, 2008 Pearson Education, Inc.

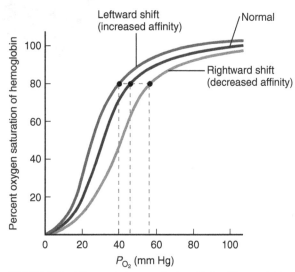

FIGURE 19.1   Effects of changes in the affinity of hemoglobin for oxygen on the hemoglobin-dissociation curve.

understand why those directional shifts would occur by examining each of the factors that affect hemoglobin's affinity for oxygen.

In the following activities, you will examine the important characteristics of the red blood cells (RBCs) (erythrocytes), which transport oxygen from the lungs to the tissues. A decreased ability to transport oxygen produces the condition called *anemia*, which can be caused by a decrease in the number or size of the red cells or by the amount of hemoglobin in the blood. To diagnose the cause of anemia accurately, the complete status of the erythrocytes must be examined—hematocrit, blood hemoglobin concentration, RBC count, RBC size, and percent hemoglobin per cell. These parameters will be measured using either your own blood or blood samples from another animal (cow, pig, rat, frog, and so on). You will also examine the movement of erythrocytes in the microcirculation (arterioles, capillaries, and venules), where oxygen is delivered to the tissues.

Blood Hematocrit

The **hematocrit** (Hct) or **packed cell volume** (PCV) is the percent volume of whole blood that is occupied by red blood cells (erythrocytes). It is determined by centrifuging the blood in special hematocrit capillary tubes. The percent of whole blood composed of erythrocytes is determined by the height of the red cells in the tube compared with the height of the total column of blood. The average normal hematocrits and their ranges for males and females are as follows:

	Average	Range
Males	46%	43–49%
Females	41%	36–45%

The hematocrit can fall to as low as 15% in severe anemia or rise to as high as 70% in **polycythemia**.

▶ ACTIVITY 19.1
Blood Hematocrit Measurement
Materials

- ☐ Disposable lancet (if using human blood)
- ☐ 5 ml whole blood
- ☐ Heparinized capillary tubes
- ☐ Sealing compound
- ☐ 1 microhematocrit centrifuge

1. Puncture your finger, using a sterile lancet to obtain a drop of blood. Wipe off the drop that forms (why?) and allow a second drop to accumulate.
2. Touch the red-circled end of a heparinized capillary tube to the drop. Hold the tube in a horizontal position and allow the blood to enter until the tube is one-half to three-fourths full.
3. Seal one end of the tube by pushing it into a tablet of sealing compound and rotating it to form a plug.
4. Place the capillary tube in a microhematocrit centrifuge with the plug end to the outside and centrifuge for 4 min.
5. At the end of 4 min, measure the height of the red cell column and the height of the cells plus the plasma (in millimeters). Calculate the hematocrit using the following formula.

$$\text{Hct }(\%) = \frac{\text{Height of red cells (mm)}}{\text{Height of red cells and plasma (mm)}} \times 100$$

Some labs employ a hematocrit "reader" that reads the hematocrit value directly on a scale. **Enter your data in the table in the Laboratory Report.** ◀

⬢ STOP AND THINK

Although humans adapt to high altitudes by increasing their hematocrit, some animals do so by having a hemoglobin molecule that shows a greater ability to bind oxygen at lower PO_2 levels. The oxygen dissociation curve is a graph that shows the percent saturation of hemoglobin at various partial pressures of oxygen. When the curve shifts to the right, oxygen dissociates from hemoglobin, whereas a shift to the left indicates that more oxygen binds to hemoglobin at the same PO_2 levels. Would you expect a llama, a mammal found in the Andes, to have an oxygen dissociation curve that shifts to the left or right when compared to a mammal living at sea level, given similar oxygen needs?

Copyright © 2015, 2011, 2008 Pearson Education, Inc.

Hemoglobin Determination

The amount of oxygen that blood can carry is closely related to the concentration of hemoglobin in the blood. Each gram of hemoglobin is capable of carrying 1.34 ml of oxygen when the hemoglobin is completely saturated. Normally, each 100 ml of blood contains approximately 15 g of hemoglobin, which is distributed among the 500 billion red cells in each 100 ml of whole blood. The red blood cell is essentially a bag of hemoglobin because as much as 34% of the red blood cell by weight is hemoglobin. The average normal concentrations of hemoglobin and their ranges for males and females are as follows:

	Average	Range
Males	15.4 g/100 ml of blood	13.6–17.2 g
Females	13.3 g/100 ml of blood	11.5–15.0 g

A concentration of less than 10 g/100 ml of blood is usually considered to indicate anemia, but major health difficulties seldom develop until a level of 7.5 g/100 ml of blood is reached.

In clinical practice, the blood hemoglobin (Hb) is usually measured by a colorimetric method such as the cyanmethemoglobin method described in this section. Other simpler tests are often performed in the laboratory to give an approximate Hb value. The values obtained by using the simpler Tallquist or Sahli methods should be checked against those yielded by the more precise cyanmethemoglobin method.

▶ ACTIVITY 19.2
Hemoglobin Determination

1. Tallquist Method

Materials

☐ Tallquist reference scale

☐ Tallquist blotting papers

☐ Disposable lancet (if using human blood)

☐ 5 ml whole blood

This test uses a book of special Tallquist blotting papers and a color comparison chart displaying different intensities of red. These intensities correspond to different concentrations of Hb found in human blood.

Obtain a drop of blood and place it on a piece of blotting paper. Before the blood dries or coagulates, match its color to the closest color on the comparison chart. The number by each color represents the percent of Hb in the blood. Multiply this number by the Tallquist standard of 16.5 g to give you grams of Hb per 100 ml of blood. **Enter your data in the table in the Laboratory Report and answer the questions that follow.**

How does this reading compare with the value obtained through the following colorimetric method? Is your Hb within the normal range for your sex?

2. Sahli Method

Materials

☐ 10 ml 0.1 M HCL

☐ Graduated Sahli tube

☐ Sahli pipette

☐ Glass rod

☐ Sahli comparator block

In this method, the blood hemoglobin is converted to a brownish hematin compound by the action of hydrochloric acid. The higher the hemoglobin concentration, the more intense the hematin color will be.

a. Place five drops of 0.1 M hydrochloric acid (HCl) in the bottom of a graduated Sahli tube. This amount should fill the tube to around the 10% mark on the scale.

b. Lance your finger to obtain a drop of blood. Place the tip of a disposable Sahli pipette in the drop and gently suction a solid column of blood into the pipette up to the 20-mm mark (0.02 ml). When suctioning, tilt the pipette horizontally and make sure that the other end is not pressed against the skin but is just immersed in the blood drop. If you draw in too much blood, touch the pipette tip to a filter paper or tissue to draw out the excess blood. Do not allow air to enter the pipette column or you will invalidate your results.

c. Insert the tip of the pipette beneath the surface of the HCl in the Sahli tube and gently blow out the blood *without placing your mouth on the tube*. Rinse the pipette of any blood by drawing the solution in and out of the pipette two times.

d. Mix the blood and HCl by stirring with a glass rod and then let the tube stand for 10 min.

e. Place the tube in the comparator block and hold it up to a strong light. Add distilled water drop by drop to the hematin solution (stir after each addition) until its color matches the color of the standard color on the comparator.

f. Read the scale on the Sahli tube to obtain the percent of Hb and grams of Hb per 100 ml

Copyright © 2015, 2011, 2008 Pearson Education, Inc.

of blood. Note that the Hb standard used in calibration can vary from tube to tube. The standard (g Hb) used is imprinted on each tube.

3. Cyanmethemoglobin Method Using a Colorimeter

Materials

- ☐ Disposable lancet (if using human blood)
- ☐ 5 ml whole blood
- ☐ 1 Spectronic 21 or other colorimeter
- ☐ 20 ml cyanmethemoglobin standard solution
- ☐ 20 ml cyanmethemoglobin reagent
- ☐ 1 sheet standard graph paper
- ☐ Sahli pipettes (# = number of unknowns)

In this activity, you will employ a very accurate colorimetric method for the determination of hemoglobin concentration in your own blood. (The colorimeter is shown in Figure 19.2) The cyanmethemoglobin test is based on the reaction of hemoglobin with *reagent* solution containing potassium cyanide (KCN) to form cyanmethemoglobin, a colored compound. The concentration of cyanmethemoglobin is determined by comparing on a **standard curve** the amount of light the compound can absorb with the amount of light absorbed by compounds with known concentrations of hemoglobin.

Place test solutions and standards in clean, matched colorimeter cuvettes. Place these into the appropriate slot of a colorimeter, using all the directions and precautions suggested by the manufacturer. In general, cuvettes should be wiped clean and dry on the outside, free of bubbles, and positioned as directed by the manufacturer. These procedures must be followed carefully to obtain accurate results.

To operate the colorimeter:

a. Turn on the instrument and let it warm up for several minutes.
b. Set the wavelength to the desired position.
c. Fill a cuvette at least ¾ full with distilled water and place it in the sample compartment. Close the lid to the sample compartment.
d. Use the 100%T dial to adjust the reading on the display to 100%T (transmittance) or 0%T (absorbance).
e. Remove your blank, replace it with your sample, and close the sample door lid. Set the MODE selector for transmittance or absorbance.
f. The reading will be displayed on the digital readout. Record your data.

For a more accurate conversion between optical density (the absorbance of an optical element for a given wavelength λ per unit distance) and percent transmittance, use Table 1 in Appendix A.

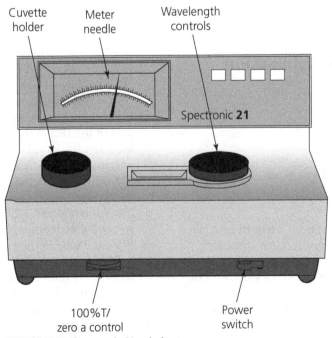

FIGURE 19.2 Spectronic 21 colorimeter.

Copyright © 2015, 2011, 2008 Pearson Education, Inc.

TABLE 19.1   Volumes of Standard Solution and Reagent Used in Setting Up Standard Curve

Volume (ml)		Hemoglobin (g %)
Standard Solution	Reagent	
0	20	4
1	16	3
2	12	2
3	8	1
4	4	0
5	0	

Setting Up the Standard Curve

Set up the standard curve by increasing dilutions of cyanmethemoglobin *standard solution;* 5.0 ml of the commercially obtained and undiluted standard corresponds to 20.0 g% hemoglobin. Dilutions must be made with the cyanmethemoglobin *reagent solution*—never with water. Prepare the dilutions using the volumes of standard and reagent shown in Table 19.1.

Transfer the dilutions to well-matched cuvettes. Set the wavelength of the colorimeter to 540 nm. Adjust the instrument so that the blank tube (0%) has zero absorbance (optical density) or 100% transmittance. Take the readings for the standard and plot percent transmission versus concentration on semilog graph paper (or absorbance [optical density] versus concentration on standard graph paper).

Determining an Unknown Sample

Place 5.0 ml of cyanmethemoglobin *reagent* in a test tube. Using a Sahli pipette, add exactly 0.02 ml of blood. Mix the contents by inverting the test tube several times. Transfer the contents to a cuvette and read the percent transmittance as compared with the reagent blank. Transfer the reading to the standard curve and obtain the hemoglobin concentration in grams percent.

3A. CYANMETHEMOGLOBIN METHOD, VERNIER® VERSION

Materials

- ☐ Vernier calorimeter
- ☐ Vernier Lab Pro
- ☐ Cuvettes
- ☐ 5 ml whole blood
- ☐ 20 ml cyanmethemoglobin standard solution
- ☐ 20 ml cyanmethemoglobin reagent
- ☐ Sahli pipettes
- ☐ 1 sheet standard graph paper

Materials needed for this exercise will have been set up for you. The colorimeter needs to warm up for about 5 min prior to collecting data. Prepare the computer for data collection by opening the file Absorb-Conc Colorimeter in the "*Experiments\ Probes & Sensors\Colorimeter*" folder.

Calibrating and Using the Vernier Colorimeter

If your colorimeter has a curved profile and a CAL button, press the < or > button on the colorimeter to select a wavelength of 565 nm (green) for this activity. Press the CAL button until the red LED begins to flash and then release it. When the LED stops flashing, the calibration is complete. Proceed to "Setting Up the Standard Curve." If your colorimeter does not have a CAL button, use the following procedure to calibrate it.

a. Set the wavelength to 565 nm.
b. Choose **Calibrate** from the experiment menu and then select **Perform now**.
c. Turn the wavelength knob on the colorimeter to the 0% T position.
d. Type 0 in the edit box under **Reading 1**.
e. When the displayed voltage stabilizes, click **Keep**.
f. Turn the knob of the colorimeter to the green LED position (565 nm).
g. Type **100** in the edit box under **Reading 2**.
h. Place a cuvette, three-fourths full of distilled water (blank), in the cuvette slot.
i. When the displayed voltage stabilizes, click **Keep**.
j. Remove the blank and place the unknown in the light path.
k. Read and record the percent transmittance and absorption.

For a more accurate conversion between optical density (the absorbance of an optical element for a given wavelength λ per unit distance) and percent transmittance, use Table 1 in Appendix A.

Setting Up the Standard Curve

The standard curve is set up by increasing dilutions of cyanmethemoglobin *standard solution*; 5.0 ml of the commercially obtained and undiluted standard corresponds to 20.0 g% hemoglobin. Dilutions must be made with the cyanmethemoglobin *reagent solution*—never with water. Prepare the dilutions, using the volumes of standard and reagent shown in Table 19.1.

Transfer the dilutions to well-matched cuvettes. Set the wavelength of the colorimeter to 565 nm. Take readings for each of the standards

Copyright © 2015, 2011, 2008 Pearson Education, Inc.

by clicking **Collect** and **Stop** before and after each reading. Record your readings and plot percent transmission versus concentration on semilog graph paper (or absorbance [optical density] versus concentration on standard graph paper).

Determining an Unknown Sample

Place 5.0 ml of cyanmethemoglobin *reagent* in a test tube. Using a Sahli pipette, add exactly 0.02 ml of blood. Mix the contents by shaking the test tube several times. Transfer the contents to a cuvette and read the percent transmittance as compared with the reagent blank. Transfer the reading to the standard curve and obtain the hemoglobin concentration in grams percent. ◄

Blood Cell Counting

Blood contains three specialized classes of cells, or formed elements: (1) red blood cells (RBCs), or erythrocytes, which transport oxygen and carbon dioxide; (2) white blood cells (WBCs), or leukocytes, which combat infections and invading organism; and (3) platelets, or thrombocytes, which prevent loss of blood. For these cells to carry out their functions properly, they must be present in sufficient numbers but not in excess. Thus, counting blood cells is an important technique because it helps establish the blood's capacity for performing these functions. The following are normal blood cell values (M = million).

Red blood cells
Males 5.4 ± 0.8 M/mm^3
Females 4.8 ± 0.6 M/mm^3

White blood cells
Males   7000–9000/mm^3
Females   5000–7000/mm^3

Platelets
150,000–400,000/mm^3
Average = 300,000/mm^3

RBCs and platelets are not true "cells" as we have come to define the term. Both lack nuclei and are unable to undergo mitosis to form daughter cells. Actually, they are nothing more than bags to carry specific chemicals: hemoglobin in the RBCs and platelet factor three in the platelets. If a gram of hemoglobin in the RBCs is maximally saturated with oxygen, it can carry about 1.34 ml of O_2. In each 100 ml of blood, there is roughly 15 g of Hb; hence, about 20 ml of oxygen can be carried in every 100 ml of blood.

CLINICAL APPLICATION

Living High and Training Low

"Live high, train low" is a mantra that identifies a method common to many elite athletes to maximize their performance levels during competition. The physiology behind this method is based on the body producing increasing amounts of the hormone *erythropoietin* in response to the lower PO_2 levels that are present at higher altitudes. Erythropoietin is a hormone secreted principally by the kidneys in response to hypoxia. It increases the production of red blood cells (the hematocrit), hence the ability to transport oxygen. "Training low" facilitates an increased ability to train due to the higher PO_2 levels at lower altitudes. This training regimen allows the cardiovascular system to adjust gradually to the increased workload associated with the increased hematocrit. Illegal and physiologically dangerous methods, such as blood doping and injecting erythropoietin into the body, have also been used to achieve this same result.

Anemia often results from an abnormal decrease in the number of erythrocytes so that insufficient oxygen is carried to the tissues and they become oxygen starved. Other factors can also cause anemia, such as decreased hemoglobin in each cell, decreased cell size, and hemorrhage.

Hemocytometer Counting Chamber and Dilution Pipettes

Although many clinics are now using automatic devices such as the Coulter counter to make their cell counts, the standard techniques are still based on the use of the hemocytometer counting chamber (Figure 19.3).

Each of the two counting chambers is 9 mm^2 and is divided into nine squares, each measuring 1 mm^2 (Figure 19.4) The four corner squares are used for counting leukocytes and are divided into 16 smaller squares to make the counting process easier. The center 1-mm^2 square is divided into 25 small squares (1/25 mm^2), and each of these is further subdivided into 16 smaller squares for counting ease. The 1/25-mm^2 squares

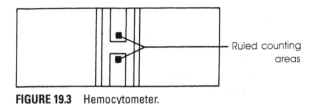

FIGURE 19.3 Hemocytometer.

Ruled counting areas

Copyright © 2015, 2011, 2008 Pearson Education, Inc.

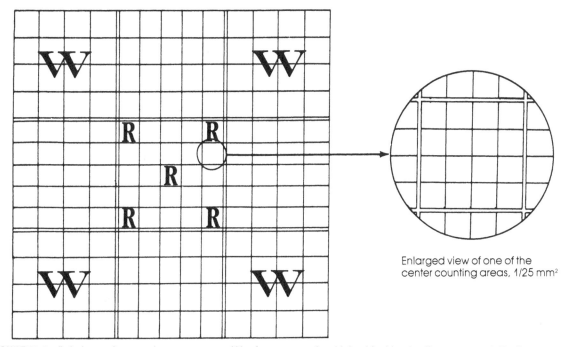

FIGURE 19.4 Ruled counting area hemocytometer. W refers to areas in which white blood cells are counted; R refers to areas in which red blood cells are counted.

are bounded by double lines and are used for counting red blood cells.

The blood cells are so numerous that they must first be diluted before they are placed in the hemocytometer for counting. Special Thoma pipettes (Figure 19.5) are usually used in making these dilutions. Alternatively, a RBC Unopette kit can be used. This kit consists of a disposable diluting pipette system that provides a convenient, precise, and accurate method for diluting the blood.

Once the dilutions are made, the hemocytometer is charged by touching the tip of the pipette to the junction of the hemocytometer and its coverslip (Figure 19.6). The diluted blood will flow in by capillary attraction to charge the counting chambers. After 2–3 min, the cells will have settled to the bottom of the chamber, and you will be ready to begin counting.

▶ **ACTIVITY 19.3**

Red Blood Cell Calculations

Materials

☐ 1 hemocytometer and cover slip

☐ 5 ml whole blood

☐ 1 Thoma red cell blood dilution pipette or Unopette RBC kit

☐ 1 compound microscope with 10× and 40× objectives

☐ 100 ml distilled water in beaker

☐ 5 ml Hayem's or Gower's solution

☐ Disposable lancets

☐ Bulb-type pipette filler

☐ 70% alcohol swabs

1. Red Blood Cell Counts

a. Obtain an RBC pipette and hemocytometer. Place the hemocytometer on the microscope stage and examine it so that you are able to

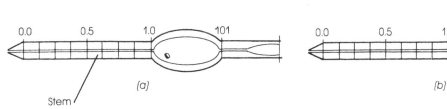

FIGURE 19.5 Thoma blood dilution pipettes. (a) Red cell pipette. (b) White cell pipette.

Copyright © 2015, 2011, 2008 Pearson Education, Inc.

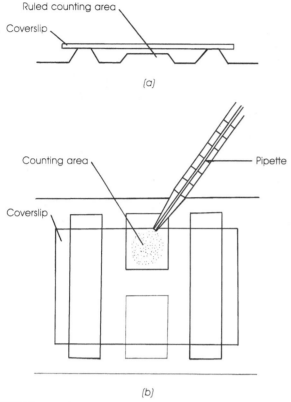

Ruled counting area

Coverslip

(a)

Counting area

Pipette

Coverslip

(b)

FIGURE 19.6 Loading the hemocytometer counting chamber. (a) Side view of counting area. (b) Procedure for loading the chamber.

identify the counting areas. Use the low power to find the center 1-mm^2 square and high power to focus on the smaller, 1/25-mm^2 squares.

b. Clean and dry the pipette *before* and *immediately after* using it, with the following solutions in the following order: distilled water (hydrogen peroxide for coagulated blood), alcohol (95%), ether or acetone.

Allow the ether to drain out by gravity from the upper end of the pipette. Then draw air through the pipette, using an aspirator. Never blow air through the pipette—this leaves water droplets in the pipette, which can alter the count.

Note: Coagulation of blood in the pipettes can be greatly reduced by drawing a heparin solution (1 mg of heparin in 500 ml of Ringer's solution) into the pipette and then blowing it out before drawing blood.

c. Clean the hemocytometer with distilled water and dry it with a soft tissue. *Do not* use an organic solvent such as alcohol, acetone, or ether because it will damage the hemocytometer counting area.

d. Place a few milliliters of Hayem's or Gower's solution in a clean watch glass. These

solutions preserve the corpuscles and prevent coagulation.

e. Clean the tip of your finger with 70% alcohol, let it dry, and then lance the finger to obtain a drop of blood. Wipe off the first drop with a tissue and allow a second drop to collect.

f. Attach a bulb-type pipette filler to the upper end of the RBC. Keeping the pipette in a horizontal position, insert the tip into the drop of blood. Compress the bulb gently and gradually release the pressure to draw blood up to *exactly* the 0.5 mark on the pipette. Be sure that the pipette tip is in the blood drop at all times so that a *solid column* of blood fills the pipette. If air bubbles enter the pipette, blow them out, clean the pipette, and try again. Keep the bulb compressed to hold the blood column intact. Wipe the excess blood from the pipette tip with a tissue. If excess blood is obtained, touch the pipette tip to a tissue to draw the blood back to the 0.5 mark.

g. Place the pipette tip in the diluting fluid and draw the fluid to exactly the 101 mark. Remove the bulb, cover the two ends of the pipette with your thumb and forefinger, and move the pipette in a circular figure-eight motion for 2 min.

h. After diluting the cells, reattach the bulb and blow out about three to five drops to remove the diluting fluid from the stem of the pipette. Place a coverslip over the counting area of the hemocytometer. Touch the tip of the pipette to the junction of the coverslip and the hemocytometer. The diluted cells will flow in by capillary attraction to charge the chamber. Allow 2 min for the cells to settle before beginning your count.

i. Focus on the center square of the counting area using high power. Count the number of red cells in five of the 1/25-mm^2 squares and take the average. Usually, the four outer squares and the middle square are counted. In your counting, you will find that some cells touch the boundary lines around the squares. Count the cells that touch on two sides of the square and omit those that touch on the other two sides.

j. Calculate the number of RBCs per cubic millimeter of blood by taking into account the following *multiplication factors*:

The blood was diluted 200 times in the pipette.

Therefore, you must multiply the average number per square by 200 (×200).

The depth of the counting chamber is 0.1 mm.

Copyright © 2015, 2011, 2008 Pearson Education, Inc.

Therefore, you must multiply the number of cells by 10 (×0).

The square counted was only 1/25 the area of the center square (1 mm^2).

Therefore, you must multiply by 25 (×25).

Multiplication factor = $200 \times 10 \times 25 = 50,000$.

For example, if you count an average of 120 RBCs per square, your RBC count is $120 \times 50,000 = 6,000,000$ RBCs/mm^3.

Record your results in the Laboratory Report. Is your RBC count within the normal range for your sex?

2. Total Oxygen-Carrying Capacity

Once the hemoglobin concentration of blood has been determined, it is possible to estimate the total milliliters of oxygen a person's blood can carry. This estimate is based on the assumption that each gram of hemoglobin is carrying a maximum of 1.34 ml of oxygen. Estimate your blood volume as follows (1 kg = 2.2 lb):

Males: 79 ml blood/kg body weight ± 10%

Females: 65 ml blood/kg body weight ± 10%

Total grams Hb in blood = Blood volume (in 100 ml) × Hemoglobin concentration (g/100 ml of blood)

Total O$_2$-carrying capacity = Total grams Hb in blood × 1.34 ml O$_2$/g Hb

3. Mean Corpuscular Hemoglobin Concentration (MCHC)

The MCHC is a measure of the hemoglobin concentration in a red blood cell. The terms *normochromic, hyperchromic,* and *hypochromic* indicate whether the MCHC is normal, high, or low. A hypochromic MCHC is associated with iron-deficiency anemia, whereas vitamin B$_{12}$ anemia shows a normochromic MCHC.

Using your hematocrit and hemoglobin values, calculate the mean corpuscular hemoglobin concentration (MCHC) for your RBCs. The normal value is 34 ± 2%.

$$MCHC(\%) = \frac{Hemoglobin\,(g/100\ ml\ blood)}{Hematocrit\,(\%)} \times 100$$

4. Mean Corpuscular Volume (MCV)

The MCV reflects the size of red blood cells. The terms *normocytic, macrocytic,* and *microcytic* indicate whether the MCV is normal, high, or low. This test will help determine the cause of anemia.

Use your RBC count and hematocrit to calculate the mean corpuscular volume (MCV) of your RBCs.

$$MCV\,(\mu m^3) = \frac{Hematocrit\,(\%RBC) \times 10^6}{RBC\ count\,(millions/mm^3)}$$

Normal = $87 \pm 2\mu^3$

Anemia can be caused by several factors such as aplastic bone marrow, RBC fragility, maturation deficiency, and hemorrhage. The determination of hematocrit, hemoglobin, RBC count, MCHC, and MCV allows one to classify the type of anemia more precisely. A microcytic MCV with a hypochromic MCHC is usually associated with iron-deficiency anemia, whereas a macrocytic MCV and a normochromic MCHC indicates a vitamin B$_{12}$–deficiency anemia. ◄

Microcirculation

The real business of the circulatory system takes place in the exchange of substances between the interstitial fluid and the small blood vessels. The collection of vessels through which this exchange occurs is often referred to as the **microcirculation**. It consists of capillaries, metarterioles, arterioles, and venules. Our current conception of this microcirculatory unit in the systemic circulation is depicted in Figure 19.7. It is often called the **Chambers-Zweifach capillary unit** after the investigators who first described its major features.

The **arterioles** contain a layer of smooth muscle in their walls that is under neural control by the autonomic nervous system (ANS). Contraction of this smooth muscle causes the lumen of the arteriole to become narrower (constrict) and thereby decrease the blood flow to that region. Relaxation of the smooth muscle causes the vessel to dilate. Thus, the ANS can divert blood from one area in the body to another in response to the overall needs of the body areas for blood. The **venules** also have a smooth muscle layer, but it is not as extensive as that around the arterioles. The **metarterioles** are direct channels between the arterioles and the venules that are always open (patent). They contain scattered amounts of smooth muscle and hence can constrict or dilate actively.

The actual exchange of substances occurs between the **true capillaries** and the interstitial fluid. The true capillaries are offshoots of either the arterioles or the metarterioles and consist of a single layer of endothelial cells and no smooth

Copyright © 2015, 2011, 2008 Pearson Education, Inc.

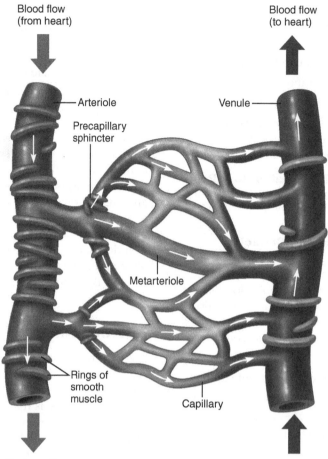

Blood flow (from heart)

Blood flow (to heart)

Arteriole

Venule

Precapillary sphincter

Metarteriole

Rings of smooth muscle

Capillary

FIGURE 19.7   Chambers-Zweifach capillary unit.

muscle. Their constriction or dilation is a purely passive process, depending on whether blood is flowing through them. At the entrance to each true capillary is a ring of smooth muscle called a **precapillary sphincter**. Its constriction or dilation controls the entrance of blood into the true capillaries. The precapillary sphincters are influenced largely by local factors resulting from tissue metabolism. Factors such as H^+ (low pH), pCO_2, temperature, histamine, and various other metabolites cause the sphincter to dilate, thus allowing blood to enter the capillaries. When the tissues are inactive and these products are not produced, the sphincters constrict and thereby shut off the blood flow through the capillaries. In this way, the tissues control their own blood flow locally—a phenomenon called **autoregulation**.

The microcirculation can be examined using several preparations, for example, the tongue, mesentery, or foot web of the frog. Because it is a simple preparation, you will use the web of the hind foot in this activity.

▶ **ACTIVITY 19.4**

Microcirculation

Materials

☐ 1 live frog

☐ 250 ml MS-222 (1%) solution

☐ Compound microscope with 10∞ objective

☐ 500-ml beaker

☐ 20 ml warm Ringer's solution (40°C)

☐ 20 ml cool Ringer's solution (10°C)

☐ 10 ml histamine, 1:10,000[1]

☐ 10 ml vasopressin or epinephrine, 1:1000[1]

☐ 10 ml lactic acid or acetic acid, 0.5%[1]

☐ Dissection equipment

1. Anesthetize the frog by double pithing or by placing the frog in a beaker containing enough 1% MS-222 (ethyl 3-aminobenzoate methanesulfonate salt) solution to cover the

[1]Most effective if the mesentery preparation is used instead of the foot.

Copyright © 2015, 2011, 2008 Pearson Education, Inc.

feet and legs. Cover the beaker and wait until the frog is anesthetized (5–6 min: corneal and withdrawal reflexes absent). Anesthesia can be maintained by covering the frog's skin with a paper towel soaked in MS-222. Return the frog to a tank containing a few inches of water, and it should recover in a few hours.

2. After anesthetizing, place the frog ventral side down on a frog board having a hole at one end. Spread the toes of one hind foot over the hole and fasten them in place loosely with pins. Do not stretch the foot too tightly or you will shut off the circulation. Keep the web moist with frog Ringer's solution and cover the frog with a wet towel to aid its respiration. Place the board on the microscope stage.

3. Examine the field under low and high power. How can you distinguish between arterioles, venules, and capillaries? Compare the rate of blood flow in each vessel. Does it flow in a pulsating manner or in a smooth flow? Estimate the diameter of each type of vessel using the RBC size (7–8 μm in diameter) as a measuring device. Notice the pliability of the RBCs as they move through the small capillaries. Look for an alteration of dilation and constriction of the capillaries over a period of several minutes. This process is called **vasomotion**. What causes vasomotion? Can you locate any leukocytes? Where are they usually found? ◄

▶ ACTIVITY 19.5

Vasoactive Agents

Apply four or five drops of each of the solutions in the following list to the frog web, washing off the web with Ringer's solution between the different solutions. The application of the solutions or sciatic nerve stimulation will disrupt your view of the web temporarily, so you will have to remember how the normal flow looks and quickly compare it with post-experimental flow. **Record your results in the Laboratory Report.**

To isolate the sciatic nerve, make a dorsal incision through the skin over the thigh muscle. Cut the fascia over the muscles and use a glass probe to separate the muscles and reach in to pull out the nerve (see Figure 15.4). Place a thread under the nerve so that the nerve can be pulled out for stimulation.

> Warm Ringer's solution (40°C)
>
> Cool ringer's solution (10°C)
>
> Histamine, 1:10,000[2]
>
> Vasopressin or epinephrine, 1:1000[2]
>
> Lactic acid or acetic acid, 0.5%[2]
>
> Light sciatic nerve stimulation
>
> Strong sciatic nerve stimulation
>
> Mechanical stimulation (draw a pin across the web) ◄

---

[2]Most effective if the mesentery preparation is used instead of the foot.

Copyright © 2015, 2011, 2008 Pearson Education, Inc.

Copyright © 2015 2011, 2008 Pearson Education

Blood Physiology I: Erythrocyte Functions

Name _____

Date _____ Section _____

Score/Grade _____

Name	Hematocrit (%)	Hemoglobin (g/100 ml)	RBC Count (m/mm³)	O₂-Carrying Capacity (ml)	MCHC (%)	CV (μm³)
Class Average	♀					
	♂					

1. What is anemia? Which blood measurements provide information on a possible anemic condition?

_____

_____

2. Briefly explain the function of the following in erythrogenesis.

Vitamin B₁₂ _____

Erythropoietin _____

Iron _____

Intrinsic factor _____

Copyright © 2015, 2011, 2008 Pearson Education, Inc.

3. Polycythemia (excess number of red blood cells) occurs in patients with chronic emphysema. Explain the mechanism responsible for this response. _____

_____

_____

4. How does hemoglobin carry both oxygen and carbon dioxide in the blood? _____

_____

_____

5. Why is the inhalation of automobile exhaust fumes life-threatening? Explain the physiology involved.

_____

_____

6. Why are hematocrits, hemoglobin concentrations, and erythrocyte counts generally lower in females than in males? _____

_____

_____

7. Where are the white blood cells in the hematocrit tube after the tube is centrifuged? _____

_____

_____

Microcirculation

1. Record your observations after each procedure, and explain the physiological action of each solution or procedure.

Experimental Procedure	Observations and Explanations
Warm Ringer's solution (40°C)	
Cool Ringer's solution (10°C)	
Histamine, 1:10,000	
Vasopressin or epinephrine, 1:1000	
Lactic acid or acetic acid, 0.5%	
Light sciatic nerve stimulation	
Strong sciatic nerve stimulation	
Mechanical stimulation	

Copyright © 2015, 2011, 2008 Pearson Education, Inc.

2. What factors are responsible for the large increase in blood flow through the skeletal muscles during exercise? _____

_____

_____

3. How is blood flow through the capillary unit controlled by central integrative centers in the medulla of the brain? _____

_____

_____

4. Outline the forces that cause movement of fluid and solutes in and out of tissue capillaries.

_____

_____

APPLY WHAT YOU KNOW

1. How does the protein content of blood affect the movement of fluid into the tissues in a capillary bed?

_____

_____

2. Changes in the hematocrit affect the movement of blood through the body. Identify how increases and decreases in the hematocrit affect the function of the heart and the ability to transport oxygen to the tissues.

_____

_____

_____

_____

Copyright © 2015, 2011, 2008 Pearson Education, Inc.

Blood Physiology II: Leukocytes, Blood Types, Hemostasis

CHAPTER 20 INCLUDES:

PhysioEx® 9.I   For more exercises on Blood Physiology, visit PhysioEx™ (www.physioex.com) and choose Exercise 11: Blood Analysis.

⚠ CAUTION!

Parts of this lab might involve working with human blood. You should handle only your own blood. Dispose of all supplies (cotton, gauze, lancets, and so on) that are exposed to blood in properly marked containers. ALL BODY FLUIDS AND SUPPLIES MUST BE TREATED AS POTENTIALLY INFECTIOUS. Read precautions for handling blood inside the front cover of this text.

OBJECTIVES

After completing this exercise, you should be able to
1. Identify and understand the functions of different white blood cells.
2. Determine a white blood cell count, using a hemocytometer.
3. Perform a differential leukocyte count, using a blood smear.
4. Determine the blood type of a sample.
5. Experimentally determine bleeding and clotting times.

Identification of White Blood Cells

In contrast to red blood cells, white blood cells (leukocytes) are nucleated and exist in several distinct types. They perform a variety of functions related to defense of the body against invading organisms.

► ACTIVITY 20.1
Identification of White Blood Cells

Materials

☐ Prepared blood slides
☐ Compound microscope with 10× and 40× objectives

Use the microscope to examine the prepared blood slides. Learn to identify each type of white blood cell (WBC) by its characteristic size, nuclear arrangement, or cytoplasmic granulation. Six types of WBCs are recognizable. The following diagrams, along with color images from your textbook or an Internet resource, will help you identify the leukocytes. ◄

Granulocytes (Polymorphonuclear Leukocytes)

Neutrophils
65% of total WBCs
10–12µm diameter
2–5-lobed nucleus
Small pink cytoplasmic granules, purple nucleus

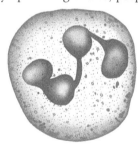

Eosinophils
2–4% of total WBCs
13-µm diameter
Bi-lobed nucleus
Coarse red-orange cytoplasmic granules, blue-purple nucleus

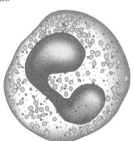

Copyright © 2015, 2011, 2008 Pearson Education, Inc.

Basophils

0.5% of total WBCs
7-μm diameter
Bi-lobed nucleus
Large, deep-blue or reddish purple cytoplasmic
granules, blue-black nucleus

Agranulocytes (Mononuclear Leukocytes)

Small Lymphocytes

25% of total WBC
7-μm diameter
Very large, spherical nucleus surrounded by thin
cytoplasm
Light blue cytoplasm (nongranular), deep blue or
purple nucleus

Large Lymphocytes

3% of total WBCs
10-μm diameter
Large oval, indented nucleus
Light-blue cytoplasm (non-granular), dark purple
nucleus

Monocytes

3–7% of total WBCs
15-μm diameter
Large kidney-shaped nucleus
Large, blue-gray cytoplasm (non-granular), blue or
purple nucleus ◀

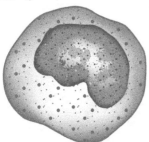

▶ ACTIVITY 20.2

White Blood Cell Counting

Materials

- ☐ 1 hemocytometer and cover slip
- ☐ 5 ml whole blood
- ☐ 1 Thoma white cell blood dilution pipette or Unopette WBC kit
- ☐ 1 compound microscope with 10× and 40× objectives
- ☐ 100 ml distilled water in beaker
- ☐ 5 ml Turk's solution
- ☐ Disposable lancets
- ☐ Bulb type pipette filler
- ☐ 70% alcohol swabs

This technique is similar to that used for counting erythrocytes (see Chapter 19, "Blood Physiology I"), with the following exceptions:

1. The WBC counting pipette that dilutes the cells 20 times is used.
2. The blood is diluted with *Turk's solution*, which consists of 1 ml of glacial acetic acid, 1 ml of aqueous 1% gentian violet solution, and 100 ml of distilled water. This acid solution lyses the red cell membranes and converts the hemoglobin to hematin. The gentian violet stains the WBCs to make identification easier.

The number of WBCs is counted in each of the four large 1-mm^2 squares in the corners of the ruled area, and the average count is determined. Use the low power for these counts.

Use the following multiplication factors to calculate the number of WBCs per cubic millimeter:

The cells were diluted 20 times (×20).
The counting chamber is 0.1 mm deep (×10).
The average number of cells in 1 mm^2 was counted (×1).
Multiplication factor = $20 \times 10 \times 1 = 200$.

How does your leukocyte count compare with the normal range of 4000–10,000/ml? What terms do we use for a deficient number of leukocytes? For an abnormally high number of leukocytes? ◀

▶ ACTIVITY 20.3

Differential Leukocyte Count

Materials

- ☐ Compound microscope with oil immersion lens
- ☐ 2 glass microscope slides
- ☐ 10 ml phosphate buffer

Copyright © 2015, 2011, 2008 Pearson Education, Inc.

☐ Wright's stain

☐ 50 ml distilled water in beaker

A determination of the **total leukocyte count** ($7500/mm^3$ average) is an important clinical measurement, but a more accurate diagnosis is obtained by making a **differential WBC count**. In a differential count, the percentage of each type of leukocyte in the total leukocyte population is determined. Each type of leukocyte performs a different function in the battle against infection, and each disease causes different responses by the WBCs. A few examples of alterations in the leukocyte population for various diseases are given in Table 20.1

Blood Smear Staining Procedure

1. Obtain a drop of blood by finger puncture. Place a small blood drop on one end of a clean glass slide (Figure 20.1).
2. Hold a second slide (the spreader) at a 45-degree angle to the first slide and move it toward the drop of blood. Allow the blood to spread along the edge of the spreader slide and then move the spreader in a smooth, fast motion to the other end of the first slide. This motion will deposit a thin, evenly spread film of blood across the slide. Allow the slide to air dry.
3. Using a medicine dropper, cover the slide with Wright's stain. Count the number of drops used. Allow the stain to stand 2 min.
4. After 2 min, add an equal number of drops of distilled water to the slide to dilute the stain. (You will obtain better results if you use a buffer solution in place of distilled water—1.63 g of KH_2PO_4 and 3.2 g of $NaHPO_4$ in 1000 ml of distilled water.) Blow gently on the slide to mix the buffer and stain. Let the slide stand for 4 min.
5. After 4 min, flush the slide gently with tap water. Dry the bottom of the slide and allow the top to air dry.

Rub finger with alcohol, then prick finger

Place drop on end of slide

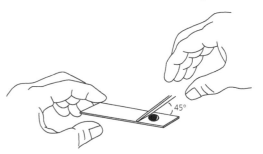

Hold two slides at 45-degree angle

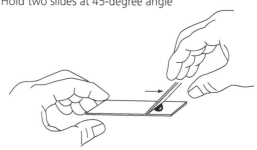

Touch drop with slide

Push back to spread drop

FIGURE 20.1 Preparation of a blood smear.

TABLE 20.1 Leukocyte Alterations Occurring with Various Diseases or Conditions

Diseases or Conditions	Symptoms
Protozoan infections, malnutrition, aplastic anemia	Neutrophilic leukopenia
Strenuous exercise, rheumatic fever, severe burns	Neutrophilic leukocytosis
Mumps, German measles, whooping cough	Lymphocytosis
Scarlet fever, parasitic infections, allergic reactions	Eosinophilia
Chronic diseases such as tuberculosis and leukemia	Monocytosis
Administration of glucocorticoid drugs	Lymphocytopenia

Copyright © 2015, 2011, 2008 Pearson Education, Inc.

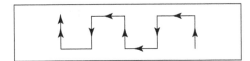

FIGURE 20.2   Scanning procedure for counting WBCs.

6. Examine the smear, first under low power and then under oil immersion, to identify the various leukocytes. If your smear is not satisfactory, prepare another one. Use a color image from your textbook or an alternate source, such as the Internet, to help you identify the different types of leukocytes.

7. Count the number of each type of WBC on the slide, recording each on a tally sheet as you identify it. Count and identify 100 WBCs and record your results in the Laboratory Report. Start scanning the bottom right of the smear and proceed using the scanning pattern shown in Figure 20.2.

8. After counting 100 cells, express the results in percentages. How do your percentages compare with the normal percentages? **Enter your observations and answer the questions in the Laboratory Report.** ◀

Blood Typing

Many clinical conditions require the transfusion of whole blood. Transfusions cannot be performed indiscriminately between persons, however, because of the possibility of antigen/antibody reactions producing **agglutination** of red cells. Agglutination refers to a clumping of red cells together. Why would agglutination be dangerous?

The human red cell has about 30 commonly occurring **antigens** on its membrane. In blood typing terminology, these are called **agglutinogens**. These agglutinogens can react with complementary **antibodies**, or **agglutinins**, in the donor's or recipient's plasma to cause agglutination of red cells.

Agglutinogens + Agglutinins → Agglutination

Although any of the 30 antigen–antibody combinations can cause agglutination, in actual practice, most agglutinations in transfusion are caused by two antigen–antibody systems—the ABO and Rh systems.

ABO System

A person can have A-type, B-type, or O-type antigens or any two of these together on the red cells. O antigens are very weak, as are the anti-O antibodies; hence they rarely cause any agglutination. For this reason, a person who has O-type blood is usually regarded as having no antigens on the red cells. Only the A and B antigens are regarded as having strong antigenicity.

Antigens are genetically determined. It should be pointed out that the ABO system is the only one in which the person's plasma automatically contains the noncomplementary antibodies to the red cell antigens. These antibodies are also determined genetically. All other antibodies found in the plasma must be formed through the entrance of an antigen into the body to stimulate antibody production.

The antigens and antibodies for each blood type are summarized in Table 20.2 along with the percentage of each type found in various races.

The differences among the races in percentage of each ABO type indicate the role of genetic determination for these blood groups.

Agglutination results from the reaction of an antigen with its complementary antibody. For example,

$$A + \alpha \rightarrow \text{Agglutination}$$

$$B + \beta \rightarrow \text{Agglutination}$$

A person with type O blood is referred to as the **universal donor**, and a person with type AB as the **universal recipient**. Explain these designations in the Laboratory Report. What are antibodies? Where are they produced in the body? What is the current theory of the mechanism of antibody production?

TABLE 20.2   Antigens and Antibodies Found in Each Blood Type and Percent Distribution of Blood Types in Various Races

Blood Type	Agglutinogen (Antigen)	Agglutinin (Antibody)	Percent Found In:		
			Caucasian	Black	Arabic
A	A	β (beta or anti-B)	43	22	5
B	B	α (alpha or anti-A)	7	29	0
AB	AB	None	3	4	0
O	None	α and β	47	45	95

Copyright © 2015, 2011, 2008 Pearson Education, Inc.

▶ ACTIVITY 20.4

Blood Typing

Materials

- ☐ Compound microscope with 10× objective
- ☐ Anti-A serum
- ☐ Anti-B serum
- ☐ Glass microscope slides
- ☐ Wooden toothpicks
- ☐ Glass marking pencil
- ☐ Disposable lancet
- ☐ Alcohol swabs

1. Obtain a clean microscope slide. Using a glass-marking pencil, mark one end A and the other end B.
2. Lance your finger to obtain blood. Place one drop of blood on each end of the marked slide.
3. Add one drop of anti-A serum to the A side. Add one drop of anti-B serum to the B side. Mix the antiserum and blood on each side with a toothpick, using a different toothpick for each side. Spread each mixture over an area of about 0.75 in diameter. Make certain you do not mix the anti-A and anti-B antisera.
4. Observe the slide for any agglutination of red cells. If agglutination occurs on side A only, you have the blood type A. If it occurs on side B only, you have type B. If a reaction occurs on both sides, you have type AB. If no reaction occurs on either side, you have type O. Explain the antigen–antibody basis for these reactions. The strength of the agglutination reaction is not the same for every person; in some cases, it might be necessary to observe the cells under the microscope to ascertain whether agglutination has actually occurred.

Use Figure 20.3 to aid in your identification of blood type.

Rh System

In 1940, Karl Landsteiner and Alexander Wiener discovered a system of antigens in the cells of the Rhesus monkey that is different from the ABO system. After producing an antiserum (antibody) against the "Rh" factor, they tested it with human RBCs and found that 85% of the human population also has this Rh factor (that is, is Rh positive). The other 15% of the population does not have this factor (is Rh negative). In contrast to antigens in the ABO system, the Rh factor is found in all body cells, not just on the erythrocytes. Actually, there are eight types of Rh agglutinogens. However, the four strongest types react with anti-Rho (anti-D) antiserum. Hence, if your blood agglutinates with anti-Rho antiserum, we say you are Rh+. If it does not, you are Rh−.

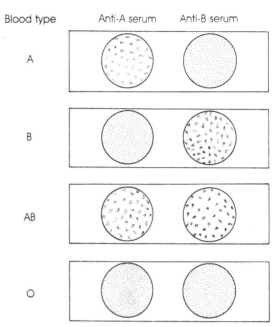

FIGURE 20.3 Antigen–antibody reactions, ABO blood type system.

⬤ STOP AND THINK

Erythroblastosis fetalis, an Rh-based hemolytic disease of the newborn, occurs only when the mother is Rh− and the child is Rh+. Can you explain why this condition manifests itself only when the mother is Rh− and not Rh+ and why it is usually not a problem when the mother is pregnant with her first child?

If a person who has Rh- blood receives a transfusion of Rh+ blood, there is usually no adverse reaction the first time, but the entrance of the Rh factor stimulates an accumulation of anti-Rh antibodies in the recipient's blood. If the same Rh− recipient receives a second transfusion of Rh+ blood, the antibodies are already present and will cause an agglutination reaction. The best known agglutination reaction of the Rh factor is **erythroblastosis fetalis**, a destruction of red cells in the newborn baby (hemolytic disease of the newborn). ◀

▶ ACTIVITY 20.5

Rh Factor

Materials

- ☐ Compound microscope with 10× objective
- ☐ Anti-Rho (anti-D) serum
- ☐ Disposable lancet
- ☐ Slide warming box

Copyright © 2015, 2011, 2008 Pearson Education, Inc.

☐ Glass microscope slides

☐ Wooden toothpicks

1. Pre-warm a clean microscope slide on a slide warming box.
2. Mix two drops of your blood with one drop of anti-Rho (anti-D) antiserum on the slide.
3. Place the slide back on the warming box and tilt the slide occasionally to aid the mixing.
4. Observe the agglutination within the first 2 min after mixing. Check under the microscope if you are in doubt. The Rh factor is usually weaker than the AB antigens, and the agglutination reaction is not as strong or as easy to detect.

Obtain the ABO and Rh blood types of all members of the class. How do the percentages of persons with each type compare with the national percentages? ◀

Blood Coagulation (Hemostasis)

Blood contains its own chemical system to **coagulate** it and thereby to prevent blood loss from the body. Coagulation is a fantastically complex process that begins as soon as blood platelets are ruptured or body tissues are damaged. In the following activity, you will examine some of the simpler processes in the coagulation mechanism.

CLINICAL APPLICATION

Hemophilia and the Blood Clotting Mechanism

Hemophilia is a group of hereditary disorders that impair the body's blood clotting mechanism. Clotting depends on the presence of several factors in the blood, which, in a series of reactions, lead to the formation of *fibrin threads*, the proteinaceous meshwork that forms the clot. Doctors diagnose hemophilia by using blood tests that determine the length of time it takes for the patients' blood to clot and the levels of clotting factors in the blood. The tests' results reveal the type and severity of hemophilia. The most common form of this condition, hemophilia A, occurs when factor VIII is absent. Hemophilia B is characterized by the absence of factor IX. Because this a sex-linked, X-chromosome disorder, more males than females suffer from the condition. People with this condition bruise easily and are subject to excessive external bleeding, internal bleeding in the joints, and, more seriously, bleeding in the brain. To treat hemophilia, patients undergo replacement therapy and are injected with concentrations of the missing or low clotting factor—either VIII or IX, depending on the condition. Due to its frequency in European royalty, hemophilia is sometimes also known as the "royal disease."

▶ ACTIVITY 20.6

Blood Coagulation

Materials

☐ Compound microscope with 10 × objective

☐ Non-heparinized capillary tubes

☐ Glass microscope slides

☐ Cover slips

☐ Wooden toothpicks

☐ Methyl violet solution

1. Bleeding Time

Clean the tip of your finger with 70% alcohol and then dry it with a piece of cotton. Puncture the finger with a lancet and record the time. At 15-sec intervals, wipe the blood drop away completely with a filter paper. (Do not touch your finger when wiping the blood away.) Continue this procedure until no more blood stains appear on the filter paper. Record this time. Calculate the bleeding time. Is it close to the normal bleeding time of 1–3 min?

2. Clotting Time

Lance your finger to obtain a large drop of blood. Note the time when the drop appears. Rapidly draw blood into a nonheparinized capillary tube by holding the tube in the drop of blood in a horizontal position.

At 30-sec intervals, break off a small piece of the capillary tube (0.5 cm) and see if clotting has occurred. Clotting has occurred when a thread of coagulated blood is visible between the two pieces of tubing.

How does your clotting time compare with the normal time of 5–8 min?

3. Observation of Fibrin Strand Formation

Place a drop of freshly drawn blood on a clean glass slide, cover it with a coverslip, and focus on it with the microscope. Place a drop of methyl violet solution to the edge of the coverslip so that the solution will run under and stain the blood. In the next few minutes, watch for the formation of fibrin strands. Use a probe or toothpick to stretch these strands to test their elasticity. ◀

Copyright © 2015, 2011, 2008 Pearson Education, Inc.

Blood Physiology II: Leukocytes, Blood Types, Hemostasis

Name _____

Date _____ Section _____

Score/Grade _____

Name	WBC Count	Differential Leukocyte Count (%)				
		Neutrophils	Eosinophils	Basophils	Lymphocytes	Monocytes
Normal values	♀6000/mm^3 ♂8000/mm^3	50–65	1–4	0–1	25–33	3–7
Class Average ♀						
♂						

1. List the major functions of the leukocytes:

Neutrophils _____

_____

_____

Eosinophils _____

_____

_____

Copyright © 2015, 2011, 2008 Pearson Education, Inc.

Basophils _____

_____

_____

Lymphocytes _____

_____

_____

Monocytes _____

_____

_____

2. Define the following terms:

Leukemia _____

_____

_____

Leukocytosis _____

_____

_____

Leukopenia _____

_____

_____

Mononucleosis _____

_____

_____

Thrombocytopenia _____

_____

_____

Hemophilia _____

_____

_____

3. Why is a person with type O blood called the "universal donor"? _____

_____

_____

4. Why is a person with type AB blood called the "universal recipient"? _____

_____

_____

5. Under what conditions is erythroblastosis fetalis possible? Why is the condition given this name?

_____

_____

6. What treatment can be given for erythroblastosis fetalis? _____

_____

_____

7. What is the difference between active and passive immunity? _____

_____

_____

Copyright © 2015, 2011, 2008 Pearson Education, Inc.

8. Identify the types of lymphocytes involved in cell-mediated and humoral immunity. Briefly explain the function of each and describe the basis for naming lymphocytes. _____

_____

_____

9. Record your values for:

Bleeding time _____ Clotting time _____

Why is bleeding time shorter than clotting time? _____

_____

_____

10. What role do the blood platelets play in coagulation? _____

_____

_____

11. Outline briefly the intrinsic and extrinsic pathways for blood coagulation. _____

_____

_____

12. What is the difference between a thrombus and an embolus? Why are they potentially dangerous?

_____

_____

APPLY WHAT YOU KNOW

1. The so-called "baby" aspirin, an 80-mg dose, is usually prescribed for people at risk of suffering a heart attack or a stroke. What is the physiological basis for this prescription?

_____

_____

_____

2. Many of the early heart transplant patients died not from rejection of the transplanted heart but from bacterial infections. How can this be explained?

_____

_____

_____

Copyright © 2015, 2011, 2008 Pearson Education, Inc.

Copyright © 2014, 2009 Pearson Education, Inc.

Physical Fitness

CHAPTER 21 INCLUDES:

 1 Vernier® Activity

 1 BIOPAC® Activity

OBJECTIVE

After completing this exercise, you should be able to

1. Experimentally estimate muscular strength, flexibility, obesity, and cardiorespiratory endurance (aerobic fitness) in a group of students.

All of us are aware of the concept of physical fitness and the decrease of fitness in recent generations, owing to the increased mechanization of our age. Yet the term *fitness* is an elusive one that means different things to different people. For our study, fitness will be defined as the capacity to meet the physical stresses encountered in life. Major components of this fitness include **muscular strength, flexibility,** and **cardiorespiratory endurance (aerobic fitness).** We will discuss **obesity** as a negative aspect because it adversely affects human health and the body systems that contribute to our overall fitness. In this chapter, you will be introduced to some simple measurements of these fitness components. It is hoped that this will spark your interest in undertaking a more thorough examination of your total fitness, using more complete and exact methods of testing. For these activities, you should wear athletic apparel (T-shirt, shorts, tennis shoes) because you will be doing moderate exercise and measuring skin folds at various sites on the body.

Muscular Strength and Endurance

Strength is one of the first things most people think of when fitness is mentioned. Muscle tone does play an important role in good posture, helping prevent lower back problems, giving us better performance in sports, and improving our figures, which gives us a definite psychological boost. An increase in muscular strength results when fast-twitch muscle fibers develop more myofilaments (actin and myosin), which provides more cross-bridges to produce tension. Exercises such as weight lifting are especially good for producing muscle hypertrophy and increased muscle strength.

Muscular endurance is the ability to contract muscles repeatedly or to sustain a single contraction. Endurance is important for many of our daily work activities and athletic endeavors. Endurance is a property of the slow-twitch muscle fibers, which increase their concentration of oxidative enzymes and capillaries with isotonic training. This allows these fibers to contract repeatedly with greatly prolonged fatigue time. It is interesting to note that increases in muscular endurance can occur with little increase in hypertrophy of the muscle fibers. In the following activity, we will examine one aspect of muscular strength and endurance by testing hand grip strength.

▶ ACTIVITY 21.1
Muscular Strength and Endurance

Materials

- ☐ 1 grip tester
- ☐ 1 stopwatch
  Or
- ☐ Materials in Activity 15.3 (Vernier® or BIOPAC® versions)

If using Vernier® or BIOPAC® systems, use the setup and calibrations described in Activity 15.3 (Chapter 15) but use the procedure described below to record your data.

1. Hold the grip tester in the center of your right palm with your arm extended at your side. Squeeze the tester as hard as possible and record the grip strength. Do this three times and use the average as your maximum grip strength. Repeat for the left hand.

2. After 1 min of rest, perform 20 consecutive grip contractions at a rate of one every 2 sec, trying

Copyright © 2015, 2011, 2008 Pearson Education, Inc.

to make each a maximal contraction. Repeat for the other hand. Record the total kilograms of force exerted for the 20 contractions and the average force of these contractions.

3. Record, on the chalkboard, the maximum grip strength, the total force for 20 contractions, and the average force of the 20 contractions for all members of the class. Calculate the mean values and ranges for the class. Compare your grip strength measurements with those of other class members of the same sex. **Enter the summarized data and answer the questions in the Laboratory Report.** ◀

Flexibility

Flexibility is the ability to move the limbs through their normal range of motion. Movement is limited by the connective tissue that covers the muscles and by the tendons that link the muscles to bone. With increasing age or inactivity, these tissues lose their elasticity, the range of motion decreases, and we become more susceptible to muscle and joint injuries. Static stretching exercises help maintain flexibility if they are performed on a daily basis, especially before and after an exercise bout. The proper technique for stretching involves a slow movement until the limit of the range of motion is reached, holding of the position for 10 sec, and then relaxation. Stretching should not be done in a jerky, rhythmic fashion, as this can damage tissues. Athletes who train their muscles (for instance, by running or power lifting) without doing stretching exercises might find their range of limb motion so limited that they appear to be muscle bound. In the following activity, we test trunk flexibility, a very important factor in preventing lower back problems.

▶ ACTIVITY 21.2
Flexibility

Materials

☐ 1 trunk flexibility tester

1. While sitting on the floor, extend your legs so that your heels touch the foot stop of the flexibility tester.
2. Place your fingers against the sliding mechanism and, with a slow, steady motion, push the slider to the farthest point you can

reach and hold for 3 sec without bending your knees. The scale reading at the slider position is your flexibility score. How does your score compare with those of the other class members? **Enter the summarized data and answer the questions in the Laboratory Report.** ◀

Body Composition

Obesity has become a critical detriment to fitness in many countries such as the United States, where approximately 80 million people are classified as obese. Obesity is an excess accumulation of fat beyond what is considered normal for the person's age and sex. Overweight does not always mean obesity if the excess weight is muscle rather than fat.

How much fat classifies a person as obese? Rough guidelines are over 20% for men and over 30% for women. The average fat percentage varies with age and sex, as shown in Table 21.1.

So what's the problem with a little excess fat? Fat is a good storage form of energy, and everyone knows that fat people are happier than the rest of the population—right? One reason fat is such a villain is that it usually compromises our muscle strength, flexibility, and cardiorespiratory endurance—the three fitness components we are measuring in the following activity. Equally important is that excess fat contributes to the development of four of our most serious health problems: cardiovascular disease, hypertension, diabetes, and cerebral vascular accident (stroke). As you can see, there are many good reasons we should shed the excess fat we are carrying around. The first step, of course, is to determine whether we

TABLE 21.1 Body Fat in Relation to Age and Sex

Age Range	Average Fat Percentage	
	Men	Women
15	12.0	21.2
18–22	12.5	25.7
23–29	14.0	29.0
30–40	16.5	30.0
40–50	21.0	32.0
Minimum	2–5	7–11

Reprinted with permission from B. Sharkey, 1990, *Physiology of Fitness*, 3rd ed. (Champaign, IL: Human Kinetics), pg. 105.

Copyright © 2015, 2011, 2008 Pearson Education, Inc.

are actually obese or just overweight with excess muscle mass.

How can we measure the relative fat and lean weight in the body? An accurate method is underwater weighing, which compares weight in air to weight underwater. This technique requires trained technicians and subjects who are comfortable underwater. Newer techniques such as dual energy X-ray absorptiometry, or DXA scanning, are now considered the standard for measuring body fat content. We will use a simpler, less expensive technique, skin fold calipers, to measure the thickness of skin and fat at representative sites around the body. These measurements are used in equations or nomograms to give an estimate of body density and percent fat. Because approximately 50% of body fat is subcutaneous, skin fold measurements can give us a fairly good estimate of fat weight. A variety of calipers are commercially available for measuring skin folds, including the inexpensive Fat-O-Meter used in this activity (available from Carolina Biological Supply Co., Burlington, NC).

▶ ACTIVITY 21.3

Body Composition

Materials

☐ Skin fold calipers

1. Acquaint yourself with the proper technique for using the calipers: Hold the skin fold between the thumb and middle finger of your left hand and the caliper in your right hand, with the scale facing you. Slide the caliper open, place it around the skin fold, and slowly close around the fold. Measure the thickness in millimeters for three trials and record the average of the two closest readings. Measurements are taken with the subject in a standing position and usually on the right side of the body.

2. Measure the appropriate skin fold sites needed for calculating the density using the Jackson & Pollock equations given below for each sex. Consult Figure 21.1 for the proper location of each site.

Men
Body Density = $1.112 - (0.00043499 \times$ sum of skin fold sites$) + (0.00000055 \times$ square of the sum of skin fold sites$) - (0.00028826 \times$ age$)$, where the skin fold sites (measured in mm) are:
S1 (Subscapula)
S2 (Triceps)
S4 (Suprailiac)
S7 (Mid Axillary)
S8 (Thigh)
S9 (Umbilicus)
S10 (Pectoral)
Women
Body Density = $1.097 - (0.00046971 \times$ sum of skin fold sites$) + (0.00000056 \times$ square of the sum of skin fold sites$) - (0.00012828 \times$ age$)$, where the skinfold sites (measured in mm) are:
S1 (Subscapula)
S2 (Triceps)
S4 (Suprailiac)
S7 (Mid Axillary)
S8 (Thigh)
S9 (Umbilicus)
S10 (Pectoral)

Copyright © 2015, 2011, 2008 Pearson Education, Inc.

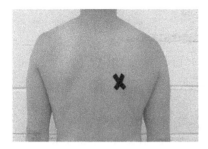
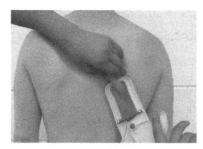

(a) SITE 1 (S1)
SUBSCAPULA
Inferior angle of the scapula,
following the natural fold of the
skin, about 1 cm below the angle

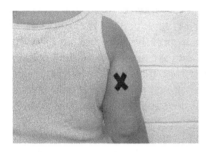
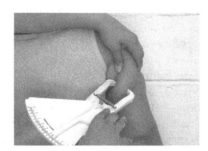

(b) SITE 2 (S2)
TRICEPS
Halfway between the acromion
process of the scapula and
olecranon process of the ulna
on the dorsum (back) of the arm

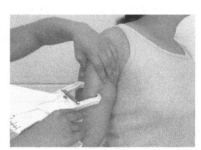

(c) SITE 3 (S3)
BICEPS
Anterior of the arm, halfway
between the greater tubercle of the
humerus and the coronoid fossa

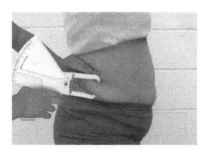

(d) SITE 4 (S4)
SUPRAILIAC
A diagonal fold immediately above
the iliac crest following the natural
fold of the skin

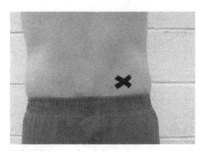
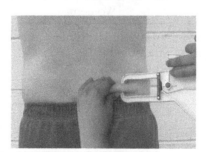

(e) SITE 5 (S5)
POST SUPRAILIAC
5 cm to the right of the first lumbar
spine

FIGURE 21.1 (a–j) Sites for measuring skin fold thickness. (The S numbers in parentheses show where measurements are inserted into the equations on page 257.)

Copyright © 2015, 2011, 2008 Pearson Education, Inc.

(f) SITE 6 (S6)
CHIN
Under the chin above the hyoid bone

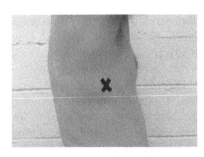

(g) SITE 7 (S7)
MID AXILLARY
Anterior diagonal fold in the mid axilla at the level of the fifth rib

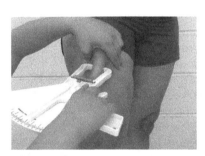

(h) SITE 8 (S8)
THIGH
Vertical fold on front of the thigh midway between the greater trochanter of the femur and the top of the patella

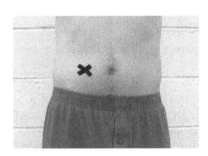

(i) SITE 9 (S9)
UMBILICUS
vertical fold to the side of the umbilicus

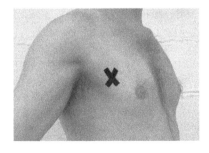
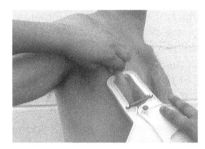

(j) SITE 10 (S10)
PECTORAL
Midway between the axillary fold and the nipple in a fold parallel to the muscle tendon.

3. Calculate your percent fat from the density, using this equation:

$$\% \text{ Fat} = \frac{457}{\text{Density}} - 414.2$$

Enter the summarized data and answer the questions in the Laboratory Report. ◀

Cardiorespiratory Endurance (Aerobic Fitness)

Aerobic fitness is the keystone of any fitness program. It is the ability of the body to use oxygen, an activity that involves the cardiovascular, respiratory, blood, and cellular enzyme systems. Aerobic

Copyright © 2015, 2011, 2008 Pearson Education, Inc.

metabolism is needed for any sustained activity that requires a high expenditure of energy. Rhythmic endurance types of exercise such as jogging, swimming, and cycling strengthen these organ systems and increase the ability to use oxygen for energy production. To achieve and maintain adequate aerobic fitness, an exercise program that requires exercise for 30 min at least five times a week at an intensity that elevates the heart rate to 70–80% of the maximum heart rate is recommended. Maximum heart rate declines with age and, although there is considerable debate about what it should be, the most acceptable current formula is

$$\text{Maximum heart rate} = 205.8 - (0.685 \times \text{age})$$

Thus, a 20-year-old college student should exercise at a training heart rate of 190 beats/min to develop good aerobic fitness.

How can we measure aerobic fitness? The best test is to measure directly the maximum oxygen a person can use per kilogram weight per minute (ml O_2/kg/min). This **maximum O_2 consumption (VO$_2$ max)** is determined while the person is exercising at peak load and rate on a treadmill or bicycle ergometer. This test requires some expensive equipment and trained technicians, which makes it impractical for testing large groups of people. Also, having an untrained person exercise at maximum load is difficult because of the potential for injury and lack of motivation for true peak performance. For comparative purposes, however, an average untrained young male will have a VO$_2$ max of about 45 ml/kg/min, whereas an elite male athlete (particularly a cyclist or cross country skier) will have a VO$_2$ max that exceeds 75 ml/kg/min. Elite female athletes in similar sports can have a VO$_2$ max that exceeds 70 ml/kg/min. Other aerobic tests use heart rate instead of O_2 consumption because, at above 120 beats/min, the heart rate and O_2 consumption increase at the same rate as the work load increases.

Aerobic fitness can be measured using either exercise heart rates or recovery heart rates because the heart rate of a trained person is lower at a particular submaximal work load and returns to the resting rate faster after exercise than the heart rate of an untrained individual.

The following activity presents two step tests that use recovery heart rate to measure cardiorespiratory fitness. The Harvard Step Test is the classic test from which other tests have been derived. It is, however, quite strenuous and discourages many people who are in the middle to lower fitness categories. For this reason, we have also included the Forest Service Fitness Test, which can be completed by older and less fit individuals.

⚠ CAUTION!

Students who have cardiovascular difficulties, such as cardiac insufficiency or hypertension, should check with their physician before taking part in these exercises.

Harvard Step Test

This widely used test was developed in the Harvard Fatigue Laboratory during World War II to screen men for physical fitness and evaluate the progress of physical training programs. The Harvard Step Test has been well validated; it measures a general endurance or physical condition that might be considered desirable for the average citizen.

The test consists of having the subject step up and down on a bench 20 in. high (16 in. for women) and then determining the heart rate during the post-exercise recovery period. It should be noted that the test was designed so that only one-third of subjects are able to complete the full 5 min of bench stepping. Thus, it is important for the subject to stop when he or she feels that he or she cannot continue. The actual time of stepping is related to the person's endurance and is part of the scoring.

▶ ACTIVITY 21.4
Aerobic Fitness I
Materials

☐ 1 20-in. (16 in. for women) stable bench or as described in the following table

1. The subject stands at attention in front of a bench 20 in. high. An observer stands behind the subject. The subject steps up and down on the bench at the rate of 30 steps (all the way up and down constitutes one step) per minute for as long as possible, up to a maximum of 5 min. The observer must be sure that the subject steps fully up on the bench without assuming any crouching position and that he or she keeps pace with the counting. If the subject is unable to keep pace for 10–15 sec, the observer stops him or her.

2. As soon as the subject stops on his or her own accord or is stopped by the examiner before or at the end of 5 min, he or she sits down. The observer notes the duration of the exercise in seconds and records the pulse from 1 to 1.5 min, from 2 to 2.5 min, and from 3 to 3.5 min after the subject has finished the exercise. The actual number of heartbeats during each of these three 30-sec periods is recorded, and the three rates are summed.

Copyright © 2015, 2011, 2008 Pearson Education, Inc.

$$Index = \frac{Duration\ of\ exercise\ in\ seconds \times 100}{2(Sum\ of\ the\ three\ pulse\ counts\ in\ recovery)}$$

The interpretation of scores is as follows:

Below 55:	Poor physical condition
55 to 64:	Low average
65 to 79:	High average
80 to 90:	Good
Above 90:	Excellent

Modified Harvard Step Test

One consistent criticism of the Harvard Step Test is the use of the same step height for all subjects. This places short persons at a disadvantage because of the larger leverage angle they must use when stepping up on a 20-in. bench. One modification that can be used to equalize the work for persons of different height is to lower the bench height for short subjects. The bench heights in the following list provide a fairer evaluation when using the Harvard Step Test. (The adjusted bench heights are for both sexes.)

Person's Height	Bench Height (in.)
<5 ft	12
5 ft to 5 ft 3 in.	14
5 ft 3 in. to 5 ft 9 in.	16
5 ft 9 in. to 6 ft	18
>6 ft	20

Forest Service Fitness Test

This test was once used to screen the physical fitness of potential firefighters for the U.S. Forest Service. It has been compared with laboratory tests of maximum O_2 consumption and found to be a valid and reliable predictor of aerobic fitness. The lower bench height does not seem to discriminate against shorter persons, and the submaximal work load does not place undue stress on the cardiovascular and respiratory systems.

The test consists of stepping up and down on a bench 15.75 in. (40 cm) high for men and 13 in. (33 cm) high for women at a rate of 22.5 steps per minute. After 5 min, the subject sits down, and a 15-sec recovery pulse count is taken from 15 to 30 sec after the test. ◄

CLINICAL APPLICATION

Weight Loss: The Preventive Key

In a country and age in which being overweight or obese has reached epidemic proportions and where medical costs to treat the obesity-related diabetes and cardiovascular conditions are skyrocketing, it is necessary to consider weight loss as an important preventive strategy. The key to weight loss is using more calories than we consume. A pound of fat (454 g) has 3500 calories of energy in it. If we consume 500 calories less each day, we should be able to lose that amount of fat in one week. Unfortunately, most diet-based weight loss strategies have been shown to be transient, with most individuals regaining the lost weight in a matter of months after stopping with the diet. Numerous studies show that individuals who diet and exercise tend to sustain their weight loss over a longer period of time. In addition to aiding in weight loss, regular exercise offers numerous other benefits such as cardiovascular fitness, better moods, lower blood pressure, and better sexual performance.

► ACTIVITY 21.5

Aerobic Fitness II

Materials

☐ 1 20-in. bench or as described in the following table
☐ 1 metronome

1. Rest for 5 min before taking the test and do not do it after strenuous physical activity, after drinking coffee or smoking, in a very warm room (over 78°F), or when anxious or excited.
2. Set a metronome or other timing device for 90 beats/min. Begin exercising to the beat of the timer with an up-up-down-down cadence of your left and right feet. You must step fully up on the bench without bending your legs, and you must keep pace with the timer.
3. After 5 min of exercise, sit down and take your pulse count for **exactly** 15 sec, starting **exactly** at 15 sec and ending **exactly** at 30 sec after exercise. The pulse can be felt on the radial artery, just below the base of the thumb, or on the carotid artery in the neck. The test is usually more accurate if a lab partner counts your pulse and checks your stepping cadence. Weigh yourself in the clothes worn during the test and record your weight.

Copyright © 2015, 2011, 2008 Pearson Education, Inc.

TABLE 21.2 Men's Fitness Scores

Post-exercise Pulse Count	Fitness Score												
45	33	33	33	33	33	32	32	32	32	32	32	32	32
44	34	34	34	34	33	33	33	33	33	33	33	33	33
43	35	35	35	34	34	34	34	34	34	34	34	34	34
42	36	35	35	35	35	35	35	35	35	35	35	34	34
41	36	36	36	36	36	36	36	36	36	36	36	35	35
40	37	37	37	37	37	37	37	37	36	36	36	36	36
39	38	38	38	38	38	38	38	38	38	38	38	37	37
38	39	39	39	39	39	39	39	39	39	39	39	38	38
37	41	40	40	40	40	40	40	40	40	40	40	39	39
36	42	42	41	41	41	41	41	41	41	41	41	40	40
35	43	43	42	42	42	42	42	42	42	42	42	42	41
34	44	44	43	43	43	43	43	43	43	43	43	43	43
33	46	45	45	45	45	45	44	44	44	44	44	44	44
32	47	47	46	46	46	46	46	46	46	46	46	46	46
31	48	48	48	47	47	47	47	47	47	47	47	47	47
30	50	49	49	49	48	48	48	48	48	48	48	48	48
29	52	51	51	51	50	50	50	50	50	50	50	50	50
28	53	53	53	53	52	52	52	52	52	52	51	51	51
27	55	55	55	54	54	54	54	54	54	53	53	53	52
26	57	57	56	56	56	56	56	56	56	55	55	54	54
25	59	59	58	58	58	58	58	58	58	56	56	55	55
24	60	60	60	60	60	60	60	59	59	58	58	57	
23	62	62	61	61	61	61	61	60	60	60	59		
22	64	64	63	63	63	63	62	62	61	61			
21	66	66	65	65	65	64	64	64	62				
20	68	68	67	67	67	66	66	65					
Body Weight	120	130	140	150	160	170	180	190	200	210	220	230	240

Reprinted with permission from B. Sharkey, 1990, *Physiology of Fitness*, 3rd ed. (Champaign, IL: Human Kinetics), pg. 296.

4. Score the test.
   a. Use your body weight and pulse count to find your fitness score in Table 21.2 (for men) or Table 21.3 (for women).
   b. Use your fitness score and age in Table 21.4 to find your age-adjusted score.
   c. Use your age-adjusted score to find your physical fitness rating in Table 21.5 (for men) or Table 21.6 (for women). ◄

Copyright © 2015, 2011, 2008 Pearson Education, Inc.

TABLE 21.3   Women's Fitness Scores

Post-exercise Pulse Count	Fitness Score											
45										29	29	29
44								30	30	30	30	30
43							31	31	31	31	31	31
42			32	32	32	32	32	32	32	32	32	32
41			33	33	33	33	33	33	33	33	33	33
40			34	34	34	34	34	34	34	34	34	34
39			35	35	35	35	35	35	35	35	35	35
38			36	36	36	36	36	36	36	36	36	36
37			37	37	37	37	37	37	37	37	37	37
36		37	38	38	38	38	38	38	38	38	38	38
35	38	38	39	39	39	39	39	39	39	39	39	39
34	39	39	40	40	40	40	40	40	40	40	40	40
33	40	40	41	41	41	41	41	41	41	41	41	41
32	41	41	42	42	42	42	42	42	42	42	42	42
31	42	42	43	43	43	43	43	43	43	43	43	43
30	43	43	44	44	44	44	44	44	44	44	44	44
29	44	44	45	45	45	45	45	45	45	45	45	45
28	45	45	46	46	46	47	47	47	47	47	47	47
27	46	46	47	48	48	49	49	49	49	49		
26	47	48	49	50	50	51	51	51	51			
25	49	50	51	52	52	53	53					
24	51	52	53	54	54	55						
23	53	54	55	56	56	57						
Body Weight	80	90	100	110	120	130	140	150	160	170	180	190

Reprinted with permission from B. Sharkey, 1990, *Physiology of Fitness*, 3rd ed. (Champaign, IL: Human Kinetics), pg. 297.

Copyright © 2015, 2011, 2008 Pearson Education, Inc.

TABLE 21.4   Age-Adjusted Scores*

Nearest Age	Enter Fitness Score																				
	30	31	32	33	34	35	36	37	38	39	40	41	42	43	44	45	46	47	48	49	50
	Age-Adjusted Score																				
15	32	33	34	35	36	37	38	39	40	41	42	43	44	45	46	47	48	49	50	51	53
20	31	32	33	34	35	36	37	38	39	40	41	42	43	44	45	46	47	48	49	50	51
25	30	31	32	33	34	35	36	37	38	39	40	41	42	43	44	45	46	47	48	49	50
30	29	30	31	32	33	34	35	36	37	38	39	40	41	42	43	44	45	46	47	48	49
35	27	28	29	31	32	33	34	35	36	37	38	39	40	41	42	43	44	45	46	47	48
40	26	27	28	30	31	32	33	34	35	36	37	38	39	40	41	42	43	44	45	46	47
45	25	26	27	29	30	31	32	33	34	35	36	37	38	39	40	41	42	43	44	45	46
50	24	25	26	28	29	30	31	32	33	34	35	36	37	38	39	40	41	42	43	44	45
55	23	24	25	27	28	29	30	31	32	33	34	35	36	37	38	39	40	40	41	42	43
60	22	23	24	25	26	27	28	30	31	32	33	34	35	36	37	37	38	39	40	41	42
65	21	22	23	24	25	26	27	28	29	30	31	32	33	34	35	36	37	38	38	39	40

Nearest Age	Enter Fitness Score																				
	51	52	53	54	55	56	57	58	59	60	61	62	63	64	65	66	67	68	69	70	71	72
	Age-Adjusted Score																				
15	54	55	56	57	58	59	60	61	62	63	64	65	66	67	68	69	70	71	72	74	75	76
20	52	53	54	55	56	57	58	59	60	61	62	63	64	65	66	67	68	69	70	71	72	73
25	51	52	53	54	55	56	57	58	59	60	61	62	63	64	65	66	67	68	69	70	71	72
30	50	51	52	53	54	55	56	57	58	59	60	61	62	63	64	65	66	67	68	69	70	71
35	49	50	51	52	53	54	55	56	57	58	59	60	60	61	62	63	64	65	66	67	68	69
40	48	49	50	51	52	53	54	55	55	56	57	58	59	60	61	62	63	64	65	66	67	68
45	47	48	49	50	51	52	52	53	54	55	56	57	58	59	60	61	62	63	64	65	65	66
50	45	46	47	48	49	50	51	52	53	53	54	55	56	57	58	58	59	61	61	62	63	64
55	44	45	46	46	47	48	49	50	51	52	53	53	54	55	56	57	58	59	59	60	61	62
60	42	43	44	45	46	46	47	48	49	50	51	51	52	53	54	55	56	57	57	58	59	60
65	41	42	42	43	44	45	46	46	47	48	49	50	50	51	52	53	54	54	55	56	57	58

Reprinted with permission from B. Sharkey, 1990, *Physiology of Fitness*, 3rd ed. (Champaign, IL: Human Kinetics), pg. 298.

*Example: If your age is 40 years and you score 50 on the step test, your age-adjusted score is 47.

Copyright © 2015, 2011, 2008 Pearson Education, Inc.

TABLE 21.5   Physical Fitness Reading—Men (Use age-adjusted score.)

Nearest Age	Superior	Excellent	Very Good	Good	Fair	Poor	Very Poor
15	57+	56–52	51–47	46–42	41–37	36–32	31–
20	56+	55–51	50–46	45–41	40–36	35–31	30–
25	55+	54–50	49–45	44–40	39–35	34–30	29–
30	54+	53–49	48–44	43–39	38–34	33–29	28–
35	53+	52–48	47–43	42–38	37–33	32–38	27–
40	52+	51–47	46–42	41–37	36–32	31–27	26–
45	51+	50–46	45–41	40–36	35–31	30–26	25–
50	50+	49–45	44–40	39–35	34–30	29–25	24–
55	49+	48–44	43–39	38–34	33–29	28–24	23–
60	48+	47–43	42–38	37–33	32–28	27–23	22–
65	47+	48–42	41–37	36–32	31–27	26–22	21–

Reprinted with permission from B. Sharkey, 1990, *Physiology of Fitness*, 3rd ed. (Champaign, IL: Human Kinetics), pg. 300.

TABLE 21.6   Physical Fitness Rating—Women (Use age-adjusted score.)

Nearest Age	Superior	Excellent	Very Good	Good	Fair	Poor	Very Poor
15	57+	56–52	51–47	46–42	41–37	36–32	31–
20	56+	55–51	50–46	45–41	40–36	35–31	30–
25	55+	54–50	49–45	44–40	39–35	34–30	29–
30	54+	53–49	48–44	43–39	38–34	33–29	28–
35	53+	52–48	47–43	42–38	37–33	32–38	27–
40	52+	51–47	46–42	41–37	36–32	31–27	26–
45	51+	50–46	45–41	40–36	35–31	30–26	25–
50	50+	49–45	44–40	39–35	34–30	29–25	24–
55	49+	48–44	43–39	38–34	33–29	28–24	23–
60	48+	47–43	42–38	37–33	32–28	27–23	22–
65	47+	48–42	41–37	36–32	31–27	26–22	21–

Reprinted with permission from B. Sharkey, 1990, *Physiology of Fitness*, 3rd ed. (Champaign, IL: Human Kinetics), pg. 300.

Copyright © 2015, 2011, 2008 Pearson Education, Inc.

Copyright 2015, 2011, 2008 Pearson Education, Inc.

Physical Fitness

Name _____

Date _____ Section _____

Score/Grade _____

Muscular Strength and Endurance

		Your Own	Females		Males	
			Mean	Range	Mean	Range
Maximum grip strength	Left Right					
Total force 20 contractions	Left Right					
Average force 20 contractions	Left Right					

Flexibility

Your score = _____

Class mean = _____ Class range = _____

Body Composition

1. Record the skin fold site and average thickness obtained:

Site					
Skin fold thickness (mm)					

Equation for age group _____ Sex _____

Density equation:

Density = _____

$$\% \text{ Fat} = \frac{457}{(\text{Density})} - 414.2$$

Body weight = _____ Lean body weight = _____ Fat weight = _____

2. Problem: Laura Wilder weighs 250 lb and has 40% body fat. She wants to lose 60 lb of fat to reduce her weight to 190 lb. What would her new percent fat be at 190 lb? _____ %

From our study of metabolic rate, we know that 2 L of oxygen is required to burn 1 g of fat and that each liter of oxygen equals 4.7 kcal of energy when fat is oxidized. Laura decides to burn off her fat by jogging at a 12-min-mile pace, which requires 200 cal/mile.

How many miles would she need to jog to burn off the 60 lb of fat, not counting any oxygen debt she might accumulate?

_____ miles

Copyright © 2015, 2011, 2008 Pearson Education, Inc.

Cardiorespiratory Endurance (Aerobic Fitness)

1. Harvard Step Test

   Duration of exercise = _____ sec

   Recovery pulse counts: 1–1.5 min = _____   2–2.5 min = _____

   3–3.5 min = _____

   Physical fitness index = $\dfrac{(\quad\quad) \times 100}{2(\quad\quad)}$ = _____

   Fitness rating = _____

2. Forest Service Fitness Test

   15-sec pulse count = _____ Body weight = _____

   Fitness score = _____ Age-adjusted score = _____

   Physical fitness rating = _____

3. Questions

   Why is a higher heart rate during the recovery period equated with a lower level of fitness?

   _____

   _____

   What is the difference between aerobic and anaerobic fitness? In which athletic events is each of these types of fitness important?

   _____

   _____

APPLY WHAT YOU KNOW

1. Based on your knowledge of human physiology, is it necessary to consume energy bars, sports drinks, and nutritional supplements to be fit? Justify your answer.

   _____

   _____

   _____

2. It is established that we live our entire lives with approximately the same number of fat cells. Given that information, are weight loss strategies involving liposuction a reasonable permanent solution to obesity? Justify your answer.

   _____

   _____

   _____

   _____

Copyright © 2015, 2011, 2008 Pearson Education, Inc.

Physiology of Exercise

OBJECTIVE

After completing this exercise, you should be able to

1. Explain the cardiovascular and respiratory systems' responses to exercise.

Of all the environmental stresses to which the body is exposed, that of exercise probably produces the greatest alteration in physiological parameters. The increased metabolic activity of skeletal muscle during exercise (a 10- to 20-fold increase) places heavy demands on the respiratory and circulatory systems and causes profound changes in other systems such as the digestive and excretory systems. Basically, these changes are brought about to satisfy the increased demand of muscle fibers for more oxygen and energy and for the removal of carbon dioxide and other metabolic waste products.

EXPLAIN THIS!

Why do muscle fibers need more oxygen if they work harder?

In the following activity, you will demonstrate several of the chief cardiovascular and respiratory responses that enable the body to supply these needs and adjust to the stresses of exercise. Such an adjustment requires integration of nearly all body systems; hence, this activity gives you the opportunity to review many of the physiological principles learned in previous chapters.

CLINICAL APPLICATION

Exercise: Benefits Beyond Weight Loss

The benefits of exercise are well documented. In addition to increasing strength and endurance, exercise also helps increase HDLs (high-density lipoproteins) and lower LDLs (low-density lipoproteins), significant factors contributing to our cardiovascular health. Additionally, blood pressure, mood, sexual activity, better sleep, and a boost in energy levels are all positively associated with regular exercise. The preventive function of regular exercise is probably most significant in that it reduces the incidence of chronic diseases such as hypertension, type II diabetes, osteoporosis, and certain types of cancer. Although exercise can assist with weight management, it is usually a reduction in caloric intake that is the most effective tool in weight loss.

Parameters Modified by Exercise

Oxygen Consumption

Oxygen consumption will be determined using the respirometer, as in the determination of basal metabolic rate (Chapter 13). The gross amount of oxygen consumed is converted to the volume of dry oxygen at standard temperature and pressure by multiplying by the STPD factor. The average individual consumes about 0.250 L of oxygen (250 ml) per minute under basal conditions. At maximal exertion, this oxygen consumption might be increased to 2.5–3 L/min (10- to 20-fold increase) in an untrained person, and in a trained athlete it might be elevated to 5 L/min (20-fold increase). The maximal oxygen uptake of a person is probably the best overall index of his or her physical fitness.

Caloric Cost

Caloric cost is the caloric energy expenditure (heat production) of the body. It is calculated by multiplying the corrected oxygen consumption (L/min) by the caloric equivalent of oxygen (4.825 kcal/L O_2 for an average mixed diet). The normal resting caloric cost is about 1.5 kcal/min and can rise to 15–20 kcal/min during exercise.

The severity of work is commonly classified on the basis of the oxygen consumption or energy expenditure, as shown in Table 22.1.

Respiratory Rate

Respiratory rate is determined by counting the deflections per minute on the respirometer record. In the resting state, the respiratory rate varies from 12 to 16 per min; it can increase to as high as 30 per min during heavy exercise.

Copyright © 2015, 2011, 2008 Pearson Education, Inc.

TABLE 22.1   Classification of Work Based on Oxygen Consumption or Energy Expenditure

Severity of Work	O_2 Consumption (L/min)	Caloric Cost (kcal/min)
Resting	0.250	1.20
Light work = O_2 consumption 2–3 × resting consumption	0.5–0.75	Up to 3.62
Medium work = 4–7 × resting consumption	1.0–1.75	Up to 8.44
Hard work = 8–12 × resting consumption	2.0–3.0	Up to 14.48
Exhaustive work = 13–20 × resting consumption	3.25–5.0	Up to 24.1

Work

The work performed during an exercise is a measure of force times distance, usually expressed as foot-pounds per minute or kilogram-meters per minute. In this activity, the exercise will consist of stepping up and down on a small bench of a designated height for a specific period. The amount of work performed is calculated as follows:

Work performed (kg-m/min) = Subject's weight (kg) × Bench height (m) × Number of steps/min

A moderate workload is considered to be about 600 kg-m/min, and a load of 1500 kg-m/min is considered a heavy one. If available, a bicycle ergometer may be used to provide a calibrated work load for the exercise.

Note: 1 kg = 2.2 lb
1 m = 39.37 in.
1 in. = 2.5 cm

Mechanical Efficiency

The efficiency of performing a certain task is simply the ratio of work done to the amount of energy used. It is possible to equate oxygen used, heat produced, and work performed through the following conversion factors:

1 L O_2 consumed = 4.825 kcal heat = 2153 kg-m work

The gross efficiency can be calculated using the following formula:

Gross efficiency (%) =
$$\frac{\text{Work performed (Kg-m)}}{O_2 \text{ used during work (L)} \times 2153 \text{ Kg-m/L}O_2} \times 100$$

Gross efficiency is somewhat misleading, however, because part of the energy used during the work period is being used just to maintain vital body activities, not to perform the work. To allow for this, use the following formula to calculate the net efficiency of the body:

Net efficiency (%) =
$$\frac{\text{Word performed (Kg-m)}}{\left(\dfrac{O_2 \text{ used} - \text{resting } O_2}{\text{during work (L)}}\right) \times 2153 \text{Kg-m/L}O_2}$$

The gross efficiency will vary from 6% to 25%, depending on the kind of work performed and the rate of doing the work. For instance, the efficiency of climbing uphill can be as high as 24%, whereas the efficiency of swimming is as low as 2–8%. In climbing uphill, the efficiency at a speed of 0.5 mph is about 6%, but at 1.5 mph, it increases to 24%. Each type of work has an optimal speed that gives the optimal efficiency for that task.

Oxygen Debt

During the performance of an exercise, a certain amount of the energy used is obtained through anaerobic metabolism. This results in an accumulation of metabolites (for instance, lactic acid) in the tissues and depletion of storage forms of energy (such as ATP and creatine phosphate). To remove these metabolites and replenish the energy stores, extra oxygen must be taken in during the recovery period following exercise. This extra oxygen is called the **oxygen debt.** There is a limit to the amount of oxygen debt a person can tolerate. An untrained person can tolerate an oxygen debt of about 10 L, a highly trained person as much as 17 L. The oxygen debt is calculated using the following formula:

Oxygen debt (L) = Total O_2 consumed in recovery period (L) −
$$\left(\dfrac{\text{Resting } O_2}{\text{consumption (L/min)}} \times \dfrac{\text{Recovery}}{\text{time (min)}}\right)$$

For the exercise performed in this activity, an oxygen debt of 1–8 L may be expected, depending on the severity of the exercise and the physical condition of the individual.

Copyright © 2015, 2011, 2008 Pearson Education, Inc.

Heart Rate

The heart rate will be determined by palpation of the carotid or radial artery or via an ECG recording. The heart rate is also a good index of the severity of the work being performed, as is shown by the values given here for work on a bicycle ergometer.

Work Load (kg-m/min)	Heart Rate (beats/min)
Resting	75
277	105
556	132
830	154
1100	177
1380	198

Generally speaking, a heart rate of less than 100 beats/min indicates light work; 100–130 beats/min, moderate work; and greater than 160 beats/min, heavy work. Usually, when the heart rate is greater than 180 beats/min, the subject is near exhaustion because the efficiency of the heart's pumping action decreases greatly at rates higher than this. Some highly trained individuals can, however, attain rates of 225 beats/min for short periods of time.

Blood Pressure

The systolic and diastolic pressures will be determined by the auscultatory method (using the stethoscope and sphygmomanometer or the Vernier® blood pressure cuff). During exercise, the systolic pressure can increase from a normal of 120 mm Hg to about 180–200 mm Hg. The diastolic pressure might increase slightly, remain the same, or even fall slightly. In general, we say that the systolic pressure reflects the force of heart contractility, whereas the diastolic pressure represents the integrity or condition of constriction of the systemic blood vessels.

Cardiac Output

The amount of blood pumped per minute by the heart is commonly measured using the Fick principle; the theory behind this method is explained in most physiology tests. It is based on calculating the volume of blood needed to transport the oxygen taken from the alveoli in a given period of time. Three measurements are required: (1) the oxygen consumption (ml/min), (2) the concentration of oxygen in the arterial blood, and (3) the concentration of oxygen in the venous blood (ml of O_2/100 ml of

blood.) The arteriovenous (AV) oxygen difference and oxygen consumption are used in the following formula to calculate cardiac output:

$$\text{Cardiac output (ml/min)} =$$
$$\frac{\text{Oxygen consumed (ml/min)}}{\text{Av oxygen difference}} \times 100$$
$$\text{(ml/100ml of blood)}$$

In this activity, you will not determine the arterial and venous oxygen concentrations directly, owing to the difficulty of obtaining blood samples. Instead, you will use what is called the modified Fick principle, in which the AV oxygen differences are obtained from the oxygen consumption values as given in the following table. Remember that this method is used only for an approximation of cardiac output in the teaching laboratory. It is not to be used in an experimental research situation.

O_2 Consumption (ml/min)	Arteriovenous O_2 Difference (ml/100 ml of blood)
250	4.5
325	4.8
400	5.0
500	5.5
600	6.0
800	6.5
1000	7.5
1200	8.3
1400	9.0
1600	9.8
1800	10.3
2000	10.9
2200	11.5
2400	12.0
2600	12.5
2800	13.0
3000	13.5
3200	13.9
3400	14.3
3600	14.6
3800	15.0

Copyright © 2015, 2011, 2008 Pearson Education, Inc.

The normal resting cardiac output is around 5 L/min. During exercise, cardiac output can rise to as high as 22 L/min in an untrained subject and to 30–40 L/min in a highly trained person.

Stroke Volume

The amount of blood forced out of the heart with each systole is around 70 ml in the average subject. During a maximal exertion, the nontrained heart will increase its stroke volume to 100–125 ml/beat, whereas the trained heart can attain values as high as 150–170 ml/beat. A stroke volume of two times the resting volume is about the maximum that can be expected during exercise.

$$\text{Stroke Volume (ml)} = \frac{\text{Cardiac output} \ (\text{ml/min})}{\text{Heart rate (beats/min)}}$$

Total Peripheral Resistance

Total peripheral resistance (TPR) is a measure of the overall resistance (constriction) of all the systemic blood vessels. It is expressed in resistance units and is calculated from our knowledge of the relationships between pressure, flow, and resistance as given in Ohm's law:

$$\text{Flow (amperes)} = \frac{\text{Pressure (volts)}}{\text{Resistance (ohms)}}$$

$$\text{Cardiac output (blood flow)} = \frac{\text{Mean blood pressure (mm Hg)}}{\text{Cardiac output (ml/sec)}}$$

$$\text{TPR (units)} = \frac{\text{Mean blood pressure (mm Hg)}}{\text{Cardiac output (ml/sec)}}$$

The mean blood pressure (BP) is a measure of the average blood pressure in the arteries over the entire cardiac cycle.

$$\text{Mean BP} = \text{Diastolic pressure} + \frac{1}{3} \ (\text{Pulse pressure})$$

The normal resting TPR is approximately 1 unit and, during exercise, it sometimes decreases to as low as 0.25 unit.

EXPLAIN THIS!

What mechanisms account for the drop in TPR during exercise? You will need to understand the factors which affect the flow of liquid in a tube to do this.

Cardiac Index

To make more accurate comparisons between the cardiac outputs of persons of different size, the cardiac index is often calculated.

$$\text{Cardiac Index} = (\text{L/min/m}^2) =$$
$$\frac{\text{Cardiac output} \ (\text{L/min})}{\text{Body surface area} \ (\text{m}^2)}$$

For humans, the cardiac index is about 3 L/min/m^2 at rest and can increase to 9–17 L/min/m^2 during exercise.

Modified Tension Time Index

The modified tension time index (MTTI) provides an estimate of the work of the heart. The work performed by myocardial tissue is proportional to the myocardial oxygen consumption, and this has been shown to be closely related to the product of heart rate (HR) and systolic blood pressure (SBP). This index is calculated as follows:

$$\text{MTTI} = \frac{\text{HR} \times \text{SBP}}{100}$$

The MTTI ranges from about 84 at rest to 360 or more during exercise.

► **ACTIVITY 22.1**
Parameters Modified by Exercise

Materials (per group)

☐ Step bench or bicycle ergometer

☐ Watch with second hand

☐ See Activity 13.1 for oxygen consumption

☐ See Activity 17.2 for measurement of blood pressure

1. The subject should come to the laboratory dressed in appropriate exercise clothes.[1]
2. The class will be divided into teams by the instructor, each team being responsible for obtaining the data on one or two parameters. The four critical measurements are oxygen consumption, heart rate, blood pressure, and work because all the other parameters are calculated from these values. **Record all your data in the Laboratory Report.**

---

[1]These tests cause cardiovascular stress, and students who have cardiovascular difficulties should not take part unless they have permission from their physicians.

Copyright © 2015, 2011, 2008 Pearson Education, Inc.

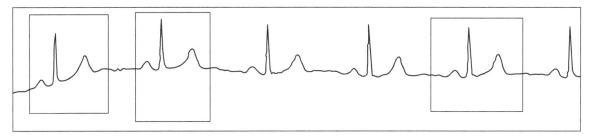

FIGURE 22.1   ECG.

3. Obtain resting values while the subject is in a standing position next to the respirometer or seated on the bicycle ergometer. **Enter these values in the table in the Laboratory Report.**

4. The subject then exercises for 2 min by stepping up and down on a bench 12–16 in. high at a rate of 20 steps per minute. The exact number of steps taken during the exercise will be counted by the team assigned to make the calculation. Heart rate and blood pressure must be taken *immediately* after the work and recorded as the work values. Note that oxygen consumption is recorded continuously through the rest, work, and recovery periods. **Enter these values in the table in the Laboratory Report.**

   If a bicycle ergometer is used, the ECG can be recorded during the last 20 sec of the work period. The work load on the bicycle should be 600–900 kg-m/min, depending on the fitness of the subject.

5. After finishing the exercise, the subject steps down from the bench and remains standing (or remains seated on the bicycle) for 8–10 min, during which time oxygen consumption, heart rate, and blood pressure will be measured and recorded each minute. **Enter these values in the table in the Laboratory Report.**
   Note: Breathing 100% oxygen often stimulates mucus production in the respiratory passageways. If the subject begins to accumulate mucus in the mouth, he or she should signal by pointing to his mouth. The operators will then close the intake valve, allow the subject to clear his mouth, place the mouthpiece back in place, and open the valve to the respirometer again. A short piece of the record will be lost, but this will not impair calculations.

6. Draw a separate slope line for each minute on the respirometer tracing and calculate the oxygen consumption for each slope.

Enter these values in the table in the Laboratory Report.

7. Place all data on the chalkboard and in the data sheet in the Laboratory Report and discuss the changes occurring in each parameter during exercise and recovery. ◄

INQUIRY-BASED ACTIVITY

Appendix C describes the format of a typical Laboratory Report. It is mandatory that you read the Appendix before you start planning your experiment.

Use any of the activities described in Chapters 17 and 18 to design your own experiment. The independent variable you choose should be easily manipulated and appropriate for the activity concerned. Please also make sure that, if using a unique measure of the dependent variables, you establish the reliability of the measuring techniques. It is recommended that you use the same measuing techniques described in the activities.

   It is quite clear, from our observations, that exercise has a significant effect on the cardiovascular system and on respiration. Because you will be testing just one independent variable in your experiment, it is important to decide how you will measure "exercise." The "dosage" of exercise (standardized in intensity and measured in units of time or standardized in time and varied in intensity) will be the independent variable. Before you begin the experiment, decide how you will ensure that you are consistent in your measurement of exercise. You may use Activities 21.4 and 21.5 to help you make the decision.

   If using the ECG as a dependent variable (as measured in Activity 17.6), understand that measurements made during the same trial run, using the same subject, should be considered one observation. In the ECG you can average the values (various intervals) measured to obtain a reading. In Figure 22.1, you can measure the suggested components (various intervals, see Figure 17.7) over three separate heartbeats and average the values to obtain one observation for the "before exercise" test and the "after exercise" tests. In other measurements, such as Blood Pressure and Heart Rate, a single measurement will suffice.

Copyright © 2015, 2011, 2008 Pearson Education, Inc.

Complete this checklist to make sure that you have covered all bases before you start your experiment:

What is the question you seek to answer?

Frame it in the form of a hypothesis.

What is the independent variable or treatment?

How will the independent variable or treatment be measured?

What is the control treatment?

How will you replicate your experiment?

How will you ensure that your subjects are similar enough to not introduce some other independent variability?

Are there any standardized variables?

What is/are the dependent variable(s)?

How will the dependent variable(s) be measured?

What are your predictions?

Construct a table to record your observations easily.

How will you present the data collected graphically?

How will you analyze the data collected?

How will you know if the differences between the treated and the untreated samples are statistically significant?

Your laboratory instructor will describe how your Laboratory Report will be written.

Copyright © 2015, 2011, 2008 Pearson Education, Inc.

Physiology of Exercise

Data sheet

Name _____

Date _____ Section _____

Score/Grade _____

Parameter Measured		Resting Values	Work Values	Recovery Values (Min in Recovery)						
				1	2	3	4	6	8	10
Oxygen consumption (L/min)										
Caloric cost (kcal/min)										
Respiratory rate (per min)										
Work (kg-m/min)										
Mechanical efficiency (%)	Gross									
	Net									
Heart rate (beats/min)										
Blood pressure (mm Hg) systolic/diastolic										
Cardiac output (L/min)										
Stroke volume (ml)										
Total peripheral resistance (units)										
Cardiac index (L/min/m^2)										
Modified tension time index										

1. Using the data you collected, compare the values for each of the following parameters, identifying the value at rest, at work, and during recovery. Explain why each parameter changes in value, providing a physiological basis for your explanation.

   a. Oxygen consumption _____
   _____
   _____

   b. Caloric cost _____
   _____
   _____

   c. Respiratory rate _____
   _____
   _____

   d. Work _____
   _____
   _____

   e. Mechanical efficiency _____
   _____
   _____

Copyright © 2015, 2011, 2008 Pearson Education, Inc.

f. Heart rate _____
_____
_____

g. Blood pressure _____
_____
_____

h. Cardiac output _____
_____
_____

i. Stroke volume _____
_____
_____

j. Total peripheral resistance _____
_____
_____

k. Cardiac index _____
_____
_____

l. Modified tension time index _____
_____
_____

2. The following diagram shows a sample respirometer record obtained during rest, work, and recovery.
Record, on the diagram, the values for oxygen consumption, heart rate, and blood pressure obtained
with your subject.

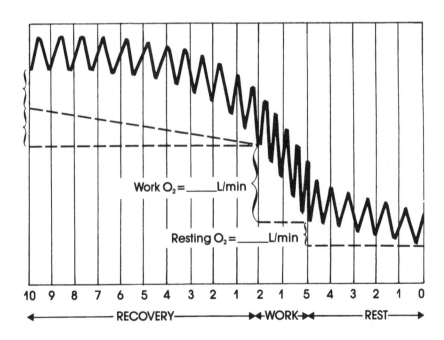

Recovery O₂ consumed

10 min O₂ debt = _____ L

Predicted rest O₂ = _____ L

Work O₂ = _____ L/min

Resting O₂ = _____ L/min

Minutes   10 9 8 7 6 5 4 3 2 1 2 1 5 4 3 2 1 0

◄————————— RECOVERY —————————► ◄—WORK—► ◄——— REST ———►

Heart rate

Blood pressure

Copyright © 2015, 2011, 2008 Pearson Education, Inc.

APPLY WHAT YOU KNOW

1. What causes the mechanical efficiency of a trained person to be higher than that of an untrained person?

   _____

   _____

2. What is responsible for the change in total peripheral resistance seen during exercise?

   _____

   _____

3. Why is the oxygen debt larger in a trained person than in an untrained person?

   _____

   _____

4. Why is the diastolic blood pressure more critical to watch during exercise and in the resting state than is the systolic blood pressure?

   _____

   _____

Copyright © 2015, 2011, 2008 Pearson Education, Inc.

Tables and Nomograms

TABLE 1 Transmission–Optical Density Table

%T	O.D.	%T	O.D.	%T	O.D.	%T	O.D.
1.0	2.000	26.0	.585	51.0	.292	76.0	.119
1.5	1.824	26.5	.577	51.5	.288	76.5	.116
2.0	1.699	27.0	.569	52.0	.284	77.0	.114
2.5	1.602	27.5	.561	52.5	.280	77.5	.111
3.0	1.523	28.0	.553	53.0	.276	78.0	.108
3.5	1.456	28.5	.545	53.5	.272	78.5	.105
4.0	1.398	29.0	.538	54.0	.268	79.0	.102
4.5	1.347	29.5	.530	54.5	.264	79.5	.100
5.0	1.301	30.0	.523	55.0	.260	80.0	.097
5.5	1.260	30.5	.516	55.5	.256	80.5	.094
6.0	1.222	31.0	.509	56.0	.252	81.0	.092
6.5	1.187	31.5	.502	56.5	.248	81.5	.089
7.0	1.155	32.0	.495	57.0	.244	82.0	.086
7.5	1.126	32.5	.488	57.5	.240	82.5	.084
8.0	1.097	33.0	.482	58.0	.237	83.0	.081
8.5	1.071	33.5	.475	58.5	.233	83.5	.078
9.0	1.046	34.0	.469	59.0	.229	84.0	.076
9.5	1.022	34.5	.462	59.5	.226	84.5	.073
10.0	1.000	35.0	.456	60.0	.222	85.0	.071
10.5	.979	35.5	.450	60.5	.218	85.5	.068
11.0	.959	36.0	.444	61.0	.215	86.0	.066
11.5	.939	36.5	.438	61.5	.211	86.5	.063
12.0	.921	37.0	.432	62.0	.208	87.0	.061
12.5	.903	37.5	.426	62.5	.204	87.5	.058
13.0	.886	38.0	.420	63.0	.201	88.0	.056
13.5	.870	38.5	.414	63.5	.197	88.5	.053
14.0	.854	39.0	.409	64.0	.194	89.0	.051
14.5	.838	39.5	.403	64.5	.191	89.5	.048
15.0	.824	40.0	.398	65.0	.187	90.0	.046
15.5	.810	40.5	.392	65.5	.184	90.5	.043
16.0	.796	41.0	.387	66.0	.181	91.0	.041
16.5	.782	41.5	.382	66.5	.177	91.5	.039
17.0	.770	42.0	.377	67.0	.174	92.0	.036
17.5	.757	42.5	.372	67.5	.171	92.5	.034
18.0	.745	43.0	.367	68.0	.168	93.0	.032
18.5	.733	43.5	.362	68.5	.164	93.5	.029
19.0	.721	44.0	.357	69.0	.161	94.0	.027
19.5	.710	44.5	.352	69.5	.158	94.5	.025
20.0	.699	45.0	.347	70.0	.155	95.0	.022
20.5	.688	45.5	.342	70.5	.152	95.5	.020
21.0	.678	46.0	.337	71.0	.149	96.0	.018
21.5	.668	46.5	.332	71.5	.146	96.5	.016
22.0	.658	47.0	.328	72.0	.143	97.0	.013
22.5	.648	47.5	.323	72.5	.140	97.5	.011

(continued)

Copyright © 2015, 2011, 2008 Pearson Education, Inc.

%T	O.D.	%T	O.D.	%T	O.D.	%T	O.D.
23.0	.638	48.0	.319	73.0	.137	98.0	.009
23.5	.629	48.5	.314	73.5	.134	98.5	.007
24.0	.620	49.0	.310	74.0	.131	99.0	.004
24.5	.611	49.5	.305	74.5	.128	99.5	.002
25.0	.602	50.0	.301	75.0	.125	100.0	.000
25.5	.594	50.5	.297	75.5	.122		

TABLE 2   Vapor Pressure of Water (mm Hg) (Values are for water in contact with its own vapor.)*

Temperature (°C)	PH_2O (mm HG)									
	0	1	2	3	4	5	6	7	8	9
0	4.6	4.9	5.3	5.7	6.1	6.5	7.0	7.5	8.0	8.6
10	9.2	9.8	10.5	11.2	12.0	12.8	13.6	14.5	15.5	16.5
20	17.5	18.7	19.8	21.1	22.4	23.8	25.2	26.7	28.3	30.0
30	31.8	33.7	35.7	37.7	39.9	42.2	44.6	47.1	49.7	52.4
40	55.3	58.3	61.5	64.8	68.3	71.9	75.7	79.6	83.7	88.0
50	92.5	97.2	102	107	113	118	124	130	136	143
60	149	156	164	171	179	188	196	205	214	224
70	234	244	255	266	277	289	301	314	327	341
80	365	370	385	401	417	434	451	469	487	506
90	526	546	567	589	611	634	658	682	707	733
100	760									

*Data from *Handbook of Chemistry and Physics*, 64th edition by Robert C. Weast (ed.). Copyright 1984 by Taylor & Francis Group LLC-Books.

Copyright © 2015, 2011, 2008 Pearson Education, Inc.

TABLE 3   Body Surface Area of Rat (M^2)*

Weight (G)	Body Surface Area (M^2)				
	0	**2**	**4**	**6**	**8**
50	0.0131	0.0134	0.0137	0.0140	0.0143
60	0.0146	0.0149	0.0152	0.0155	0.0158
70	0.0161	0.0164	0.0167	0.0169	0.0171
80	0.0174	0.0176	0.0179	0.0181	0.0184
90	0.0186	0.0189	0.0192	0.0194	0.0196
100	0.0199	0.0201	0.0204	0.0206	0.0208
110	0.0210	0.0213	0.0215	0.0218	0.0220
120	0.0222	0.0224	0.0227	0.0229	0.0231
130	0.0233	0.0235	0.0237	0.0239	0.0241
140	0.0243	0.0245	0.0247	0.0249	0.0251
150	0.0253	0.0255	0.0257	0.0259	0.0261
160	0.0263	0.0265	0.0267	0.0269	0.0271
170	0.0273	0.0275	0.0277	0.0279	0.0281
180	0.0283	0.0285	0.0287	0.0289	0.0291
190	0.0293	0.0294	0.0295	0.0297	0.0299
200	0.0301	0.0303	0.0304	0.0306	0.0308
210	0.0310	0.0312	0.0313	0.0315	0.0317
220	0.0319	0.0321	0.0322	0.0324	0.0326
230	0.0328	0.0329	0.0331	0.0333	0.0334
240	0.0336	0.0338	0.0339	0.0341	0.0343
250	0.0345	0.0346	0.0348	0.0349	0.0351
260	0.0353	0.0354	0.0356	0.0357	0.0359
270	0.0361	0.0362	0.0364	0.0365	0.0367
280	0.0369	0.0370	0.0372	0.0373	0.0375
290	0.0377	0.0378	0.0380	0.0381	0.0383
300	0.0384	0.0386	0.0387	0.0389	0.0390
310	0.0392	0.0393	0.0395	0.0396	0.0398
320	0.0399	0.0401	0.0402	0.0404	0.0405
330	0.0407	0.0408	0.0410	0.0411	0.0413
340	0.0414	0.0416	0.0417	0.0419	0.0420
350	0.0421	0.0423	0.0424	0.0426	0.0427
360	0.0428	0.0430	0.0431	0.0433	0.0434
370	0.0435	0.0437	0.0438	0.0440	0.0441
380	0.0442	0.0444	0.0445	0.0447	0.0448
390	0.0449	0.0451	0.0452	0.0454	0.0455
400	0.0456	0.0458	0.0459	0.0461	0.0462
410	0.0463	0.0465	0.0466	0.0468	0.0469
420	0.0470	0.0472	0.0473	0.0475	0.0475

*Data from A. B. Taylor and F. Sargent II, *Elementary Human Physiology: Laboratory and Demonstration Manual* (Minneapolis: Burgess, 1962).

Copyright © 2015, 2011, 2008 Pearson Education, Inc.

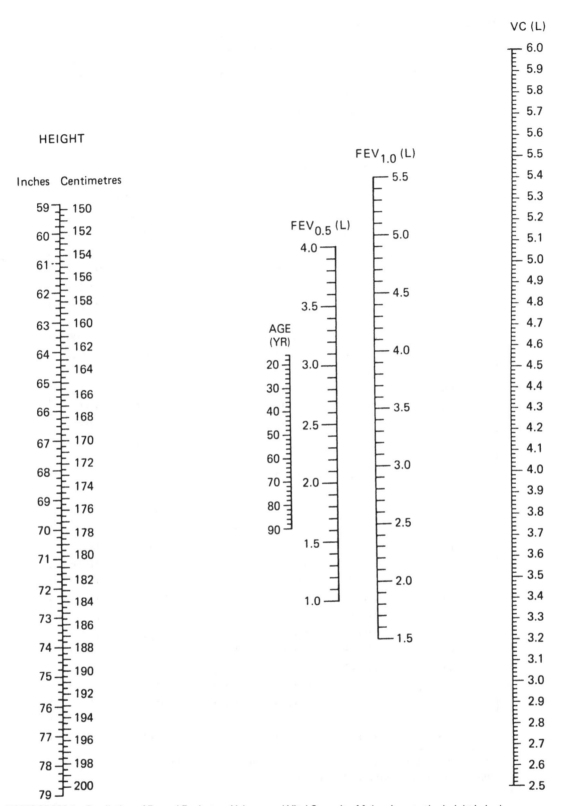

HEIGHT

Inches Centimetres

VC (L)

FEV$_{1.0}$ (L)

FEV$_{0.5}$ (L)

AGE
(YR)

NOMOGRAM 1   Prediction of Forced Expiratory Volume and Vital Capacity, Males. Locate the height in inches (or centimeters) and the age in years. Place a straightedge between these two points; the intersections will give the predicted forced expiratory volume (FEV) and the predicted vital capacity (VC).

Copyright © 2015, 2011, 2008 Pearson Education, Inc.

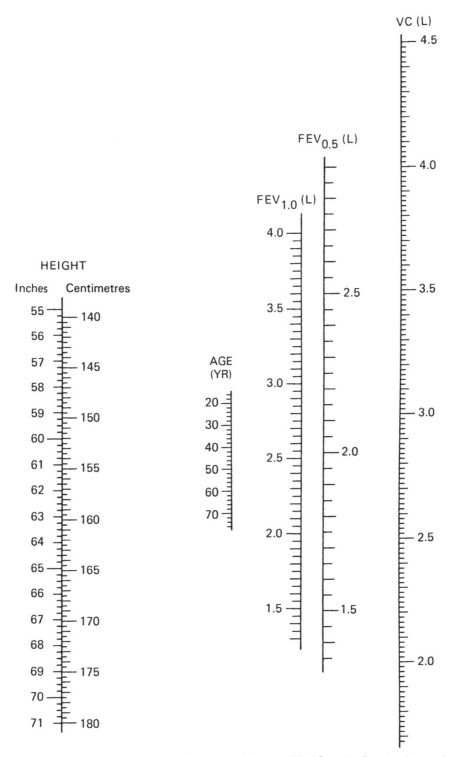

NOMOGRAM 2   Prediction of Forced Expiratory Volume and Vital Capacity, Females. Locate the height in inches (or centimeters) and the age in years. Place a straightedge between these two points; the intersections will give the predicted forced expiratory volume (FEV) and the predicted vital capacity (VC).

Copyright © 2015, 2011, 2008 Pearson Education, Inc.

Solutions

1. **Physiological Saline (Mammalian—Endotherms)**

   0.9% NaCl in distilled water

2. **Physiological Saline (Amphibian—Ectotherms)**

   0.7% NaCl in distilled water

3. **Frog Ringer's Solution**

   NaCl   6.50 g
   KCl   0.14 g
   $CaCl_2$ 0.12 g
   $NaHCO_3$ 0.20 g
   NaH_2PO_4 0.01 g
   Distilled water of a sufficient quantity (q.s.) to 1 L

4. **Mammalian Ringer's Solution**

   NaCl   9.00 g
   KCl   0.42 g
   $CaCl_2$ 0.24 g
   $NaHCO_3$ 0.20 g
   Distilled water q.s. to 1 L

5. **Locke's Solution**

   NaCl   9.00 g
   KCl   0.42 g
   $CaCl_2 \cdot 2H_2O$ 0.24 g
   $NaHCO_3$ 0.20 g
   Glucose   1.00 g
   Distilled water q.s. to 1 L (pH should be 7.6–7.8)

6. **Heparinized Mammalian Ringer's Solution**

   Add 100 units (1 mg) of heparin per 500 ml of Ringer's solution.

7. **Anticoagulants for Blood**

   Use 20 mg of sodium oxalate or 200 mg of sodium citrate per 10 ml of blood.

8. **Hayem's Solution**

   Mercury bichloride   0.5 g
   Sodium chloride   1.0 g
   Sodium sulfate   5.0 g
   Distilled water   200 ml
   Filter the solution.

9. **Gower's Solution**

   Glacial acetic acid   16.65 ml
   Anhydrous Na_2SO_4 6.25 g
   Distilled water   100.00 ml (exactly)

10. **Turk's Solution**

    Glacial acetic acid   1.0 ml
    1% aqueous solution of Gentian violet   1.0 ml
    Distilled water   100.0 ml (filtered frequently)

11. **Wright's Stain**

    Obtain from laboratory supply houses.

12. **Buffer Solution for Use with Wright's Stain**

    KH_2PO_4 1.63 g
    Na_2HPO_4 3.2 g
    Distilled water   1000 ml

13. **Benedict's Solution**

    Copper sulfate   17.3 g
    Sodium citrate   173.0 g
    Sodium carbonate (anhydrous)   100.0 g
    Distilled water q.s. to 1000 ml
    Dissolve the citrate and carbonate in 800 ml of water and filter. Dissolve the copper sulfate in 100 ml of water and then pour it slowly into the first solution while stirring constantly. Cool the solution and add water to make 1 L.

14. **Krebs Manometer Fluid**

    Anhydrous NaBr   44 g
    Triton X-100 (Rohm & Haas Co.)   0.3 g
    Evans blue   0.3 g
    Distilled water   1000 ml

15. **Brodie's Manometer Fluid**

    NaCl   23 g
    Sodium choleate (Merck)   5 g
    Evans blue   0.1 g
    Distilled water   500 ml

16. **Lugol's Solution**

    Iodine   5 g
    Potassium iodide   10 g
    Distilled water q.s. to 100 ml

Copyright © 2015, 2011, 2008 Pearson Education, Inc.

17. Phosphate Buffer

Stock solutions:
M/15 dibasic sodium phosphate 9.465 g in 1000 ml distilled water solution
M/15 sodium acid phosphate 8.0 g in 1000 ml distilled water solution
For a pH of 7, mix:
60 ml stock dibasic sodium phosphate
40 ml stock sodium acid phosphate

18. Synthetic Urine

Stock solution:
Sodium chloride   15.0 g
Potassium chloride   9.0 g
Sodium phosphate monobasic   9.6 g
Urea   36.4 g
Distilled water   1.5 L
Mix until the solution is clear. Add distilled water to bring the volume to 2 L. Check the pH level to ensure that it is within the 5–7 range for normal urine. If needed, lower the pH with 1 N hydrochloric acid or raise it with 1 N sodium hydroxide. Place a urinometer in the solution and dilute with distilled water until the specific gravity is within the 1.015–1.025 range. Add yellow food coloring for desired color.

Use the modifications below to mimic some common pathologies.

Hematuria:
Add one drop of citrated sheep blood to every 250 ml of solution.

Proteinuria:
Add bovine albumin to stock solution until the urine gives a positive test for proteins.

Diabetes mellitus:
Add glucose, bovine albumin, and acetoacidic acid to stock solution until a positive glucose, protein, and ketones test are obtained.

Diabetes insipidus:
Add distilled water to stock urine until a significant color change occurs.

19. Rat anesthesia cocktail

Ketamine is a Drug Enforcement Administration (DEA) Schedule III drug and must be kept in a double-locked area with appropriate records of administration.
Ketamine HCl (100 mg/ml)   5 ml
Xylazine HCl (100 mg/ml)   0.5 ml
Inject 0.07 ml/100 g body weight of this mixture intraperitoneally (IP).

Copyright © 2015, 2011, 2008 Pearson Education, Inc.

Inquiry-Based Activities

The scientific method calls for the use of experiments to answer questions in a manner that will either confirm or falsify the question under consideration. Typically a scientific question is framed in the form of a **hypothesis.** A hypothesis is not simply a guess but is a research-based educated guess and is usually based upon previous peer-evaluated, published information. In conducting an experiment, an investigator will be measuring **variables.** These are measurements that change during the experiment. If they change by design, then they are known as **independent variables.** Measurements that change in response to changes made in the independent variable are known as **dependent variables.** If there are other variables that can affect the outcome of the experiment, an investigator will keep those constant, and it/they will be identified as **standardized variables.** In the discussion that follows, we will use the framework of a specific experiment to help you understand how an experiment is conceived and carried out.

Suppose you are interested in investigating the effects of exercise on the functioning of the cardiovascular system. In order to perform a successful experiment one must, *a priori,* decide how it will be performed. Therefore we must formulate a question, perhaps one as simple as "What are the effects of exercise on the human body?," and then refine the question based upon our previous knowledge. In this case, it may be quite clear, from our observations and readings, that exercise has a significant effect on the cardiovascular system, causing its activity to increase. Given that, we can formulate a hypothesis more precisely and state: "Exercise increases the functioning of the cardiovascular system."

Because the investigator will be able to manipulate the amount of exercise, it will be considered the **independent variable.** Only one independent variable can be investigated in an experiment. In some instances, the experiment may consist of examining the response of the body to the presence or absence of the independent variable. In other instances, the level of the independent variable (or treatment) may be varied. The differences in the level of the treatments applied are usually based upon a prior knowledge of previous research. You can think of these levels of treatment as dosages.

If you were testing the effects of exercise on cardiovascular function, the "dosage" of exercise (standardized in intensity and varied in amount of time the exercise is performed or standardized in time and varied in the intensity level) will be the independent variable.

Because your question deals with the effects of exercise on the cardiovascular system, let us start by asking "How do you measure the functioning of the cardiovascular system?" An obvious measure of cardiovascular function would be the heart rate. In an experiment, this will be the **dependent variable.** There may be more than one dependent variable measured during an experiment. We can therefore refine our dependent variables, making them more specific. Use your knowledge and your textbook to list how else you would measure the functioning of the cardiovascular system.

Other measurements of cardiovascular function:

_____

_____

_____

Scientists will also use a **control treatment** to ensure that changes observed in the dependent variable are due to variations in the independent variable. In the example we have used, a control treatment would be "no exercise" or "before exercise." In this instance, if we were measuring heart rate as the dependent variable, a basal heart rate would be recorded before any exercise had taken place. Using controls allows an investigator to compare the results of a "treated" versus an "untreated" test subject to allow him or her to conclude that changes in the dependent variable were due to the treatments used during the experiment. In many medically related experiments, **placebos** are used as a control to ensure that any improvement observed in a patient is due to a treatment rather than the perception of a treatment.

The **methods** used in an experiment are the mechanisms used to measure changes seen in the dependent variables due to the treatments used. In

Copyright © 2015, 2011, 2008 Pearson Education, Inc.

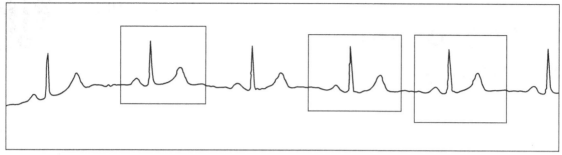

FIGURE 1 ECG strip with 3 "heart-beats" selected.

the example we have chosen, these may include describing how the investigator would count the heart rate. It is important that the description of the methods be complete enough to give subsequent investigators sufficient information to replicate the experiment accurately. In the experiment we are using, the heart rate could be measured immediately after the completion of the treatment for a period of 15 sec. This value would be multiplied by 4 to obtain a heart rate expressed in "beats per minute" (BPM).

In order to establish that the experiment performed and the results obtained are not a fluke, a scientist will **replicate** the experiments. This means performing the experiment a number of times using the same experimental setup to establish that the results obtained are consistent. Because the subjects of the experiment will tend to show variation, it would not be surprising to observe differences amongst subjects. However, the pattern of response should typically remain consistent, if the changes observed were in response to changes in the independent variable. Because using just a few subjects may still not allow us to draw significant statistical inference, the **sample size** used for the experiments is usually a very important part of the experimental design. In a physiology teaching laboratory, it may be almost impossible to accommodate the sometimes stringent requirements needed to draw statistical inference. Your instructor will advise you accordingly, as that aspect of experimental design is beyond the scope of this manual.

Once your experimental protocol has been completed, you can start collecting data. A **data table**

is usually used to organize your data collection. Many digital data acquisition systems automate data collection, along with automatically generating a graphical representation of the data collected. An electrocardiograph (ECG) machine is an example of such a device.

Some Suggestions

Measurements made during the same trial run, using the same subject, should be considered one observation. In some cases, such as in the ECG, for example, you can average the values measured (various intervals) to record data. If the ECG were being measured before and after exercise and we were examining various peaks and intervals (as described in Chapter 17), the suggested components can be measured and averaged over three or more separate heartbeats (see Figure 1) to obtain one observation for the "before exercise" test and the same for the "after exercise" test.

In other situations, such as blood pressure or heart rate, a single measurement would suffice and would in all probability be a more accurate measure, rather than three separate sequential measures made after one treatment.

● STOP AND THINK

Why would one measure be a better estimate than an average of three separate, sequential measurements of blood pressure or heart rate in this situation?

A sample data table, showing three separate dependent variables, may resemble Table 1 if you

TABLE 1 Sample Data Table Showing Three Separate Dependent Variables

Student	Before Exercise (the "Control")			After 2 Minutes of Exercise			After 5 Minutes of Exercise		
	Heart Rate	Systolic BP	Diastolic BP	Heart Rate	Systolic BP	Diastolic BP	Heart Rate	Systolic BP	Diastolic BP
A									
B									
C									

Copyright © 2015, 2011, 2008 Pearson Education, Inc.

TABLE 2 A Table Showing Data Collected During an Experiment Investigating the Effect of Exercise on the Heart Rate

Time (minutes)	Heart Rate
0	68
1	72
2	78
3	85
4	90
5	100
6	120
7	150
9	152
10	153

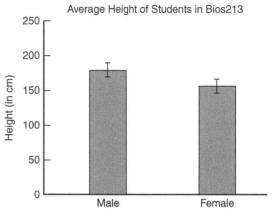

FIGURE 3   A bar graph showing the average heights of male and female students.

are investigating the effects of exercise on heart rate and blood pressure.

While a table may adequately reflect the data collected and upon scrutiny reveal some pattern of change, this may not be apparent in some large data sets. Compare the data presented in Table 2 and the graph in Figure 2.

Quite clearly, the pattern of response is more obvious in the graph rather than in the table. This becomes even more evident in data sets involving many more points over a longer period of time.

In order to graphically represent the results of your experiment, you collect the data in the form of a table (Table 1). Most physiological experiments will represent the data in either a line or a scatter-plot type of graph (Figure 2). In this case presented, the independent variable is time, and the

parameter(s) being measured (heart rate) will be the dependent variable. Other examples of these types of experiments are those involving changes in blood pressure, respiration rate, or recovery time over a period of time. When the independent variable is a categorical variable (gender, blood type, dosage of a drug, etc.), a scientist will typically use a bar graph (Figure 3) to represent the data collected.

Here are some simple rules to follow when creating a graph:

1. Give your graph a descriptive title.
2. Label your axes appropriately, clearly indicating the units being used.
3. The independent variable (time, gender, dosage) will always be plotted on the *x*-axis (horizontal) and the dependent variable (blood pressure, heart rate, membrane potential) will always be plotted on the *y*-axis (vertical).
4. In some cases, a key will be necessary to explain colors or symbols used.
5. Make sure you examine your data to evaluate the range of values. Each axis will be scaled to accommodate this range. Do not waste space. Remember, the units used on each axis will be different, so adjust your scale accordingly.
6. It is *not* always necessary to use a line to connect points on a graph.

Most modern computer-based physiological equipment will usually collect data and generate the graph automatically. Other stand-alone pieces of equipment, such as oscilloscopes, are essentially sophisticated real-time graphing machines.

Of course, being scientists, we need to demonstrate that these differences are statistically significant. A rather simple test of statistically

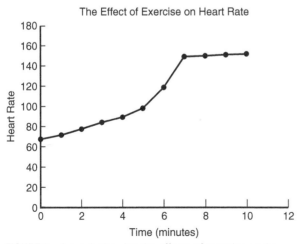

FIGURE 2   A graph showing the effects of exercise on the heart rate.

Copyright © 2015, 2011, 2008 Pearson Education, Inc.

significant differences in a comparison of treated versus untreated samples in an experiment is the **Student's T-test.**

Because the same subjects are being compared before and after periods of exercise, we can use the Student's T-Test to see if there are statistical differences between the "Before exercise/2 min exercise" values, and the "Before exercise/5 min exercise" values. Use the website www.studentsttest.com/ to calculate your statistics. Use the "*groups are matched*" and the 1-tail t-test to check for significant differences between the "before" and "after" values. Report your results using a printout of the page, the tables used to collect data, and whether your results indicated significant differences in the before and after exercise data. A p < .05 will be used to indicate a significant result.

The structure and content of your laboratory report will vary according to your instructor.

In conclusion, here are some simple guidelines to initiate the process of designing your own experiment:

1. What is the question you seek to answer? Frame it in the form of a hypothesis.

2. What is the independent variable or treatment?

3. How will the independent variable or treatment be measured?

4. What is the control treatment?

5. How will you replicate your experiment?

6. How will you ensure that your subjects are similar enough to not introduce some other independent variability? Are there any standardized variables?

7. What is/are the dependent variable(s)?

8. How will the dependent variable(s) be measured?

9. What are your predictions?

10. Construct a table to record your observations easily.

11. How will you present the data collected graphically?

12. How will you analyze the data collected?

13. How will you know if the differences between the treated and the untreated samples are statistically significant?

Copyright © 2015, 2011, 2008 Pearson Education, Inc.

Credits

Cover

Science Picture Co./Corbis.

Chapter 1

Figure 1.4. Image courtesy of ADInstruments Inc.

Chapter 3

Figure 3.1. Stanfield, Cindy L.; Germann, William J., *Principles of Human Physiology*, 3rd Ed., © 2008, pg. 517. Reprinted and electronically reproduced by permission of Pearson Education, Inc., Upper Saddle River, New Jersey.

Chapter 5

Figure 5.1. Stanfield, Cindy L.; Germann, William J., *Principles of Human Physiology*, 3rd Ed., © 2008, pg. 182. Reprinted and electronically reproduced by permission of Pearson Education, Inc., Upper Saddle River, New Jersey.

Figure 5.5. Image courtesy of ADInstruments Inc.

Figure 5.6. Image courtesy of ADInstruments Inc.

Chapter 8

Figure 8.3. Wood, Michael G. *Laboratory Manual for Anatomy & Physiology, Main Version*, 3rd Ed., © 2006, pg. 442. Reprinted and electronically reproduced by permission of Pearson Education, Inc., Upper Saddle River, New Jersey.

Figure 8.7. Stanfield, Cindy L.; Germann, William J., *Principles of Human Physiology*, 3rd Ed., © 2008, pg. 270. Reprinted and electronically reproduced by permission of Pearson Education, Inc., Upper Saddle River, New Jersey.

Chapter 10

Figure 10.1. Stanfield, Cindy L.; Germann, William J., *Principles of Human Physiology*, 3rd Ed., © 2008, pg. 571. Reprinted and electronically reproduced by permission of Pearson Education, Inc., Upper Saddle River, New Jersey.

Chapter 12

Figure 12.3. Dmitry Lobanov/Shutterstock.

Chapter 15

Figure 15.8. Image courtesy of ADInstruments Inc.

Figure 15.9. Image courtesy of ADInstruments Inc.

Figure 15.10. Image courtesy of ADInstruments Inc.

Figure 15.13. © Vernier Software & Technology. Used with permission.

Figure 15.14. Screencap from BIOPAC.

Chapter 16

Figure 16.13. Image courtesy of ADInstruments Inc.

Chapter 17

Figure 17.3. Screencap from BIOPAC.

Figure 17.9. Screencap from BIOPAC.

Figure 17.10. Screencap from BIOPAC.

Chapter 18

Figure 18.1. Based on Cindy L. Stanfield, *Principles of Human Physiology*, 5th Ed. Figure 16.10, pg. 460, Pearson Education.

Figure 18.2. Based on Cindy L. Stanfield, *Principles of Human Physiology*, 5th Ed. Figure 16.11, pg. 461, Pearson Education.

Figure 18.6. Screencap from BIOPAC.

Figure 18.7. Screencap from BIOPAC.

Figure 18.9. © Vernier Software & Technology. Used with permission.

Figure 18.10. Screencap from BIOPAC.

Figure 18.11. Screencap from BIOPAC.

Figure 18.12. Screencap from BIOPAC.

Figure 18.13. Screencap from BIOPAC.

Chapter 19

Figure 19.1. Based on Cindy L. Stanfield, *Principles of Human Physiology*, 5th Ed. Figure 17.9, pg. 485, Pearson Education.

Chapter 21

Table 21.1. Reprinted with permission from B. Sharkey, 1990, *Physiology of Fitness*, 3rd ed. (Champaign, IL: Human Kinetics), pg. 105.

Table 21.2. Reprinted with permission from B. Sharkey, 1990, *Physiology of Fitness*, 3rd ed. (Champaign, IL: Human Kinetics), pg. 296.

Copyright © 2015, 2011, 2008 Pearson Education, Inc.

Table 21.3. Reprinted with permission from B. Sharkey, 1990, *Physiology of Fitness*, 3rd ed. (Champaign, IL: Human Kinetics), pg. 297.

Table 21.4. Reprinted with permission from B. Sharkey, 1990, *Physiology of Fitness*, 3rd ed. (Champaign, IL: Human Kinetics), pg. 298.

Table 21.5. Reprinted with permission from B. Sharkey, 1990, *Physiology of Fitness*, 3rd ed. (Champaign, IL: Human Kinetics), pg. 300.

Table 21.6. Reprinted with permission from B. Sharkey, 1990, *Physiology of Fitness*, 3rd ed. (Champaign, IL: Human Kinetics), pg. 300.

Copyright © 2015, 2011, 2008 Pearson Education, Inc.

Index

Copyright © 2015, 2011, 2008 Pearson Education, Inc.

Copyright © 2015, 2011, 2008 Pearson Education, Inc.